ACS SYMPOSIUM SERIES **431**

Expert Systems for Environmental Applications

Judith M. Hushon, EDITOR
Roy F. Weston, Inc.

Developed from a symposium sponsored
by the Division of Chemical Information
at the 198th National Meeting
of the American Chemical Society,
Miami Beach, Florida
September 10–15, 1989

American Chemical Society, Washington, DC 1990

Library of Congress Cataloging-in-Publication Data

Expert systems for environmental applications
  Judith M. Hushon, editor.

    p.   cm.—(ACS Symposium series, ISSN 0097–6156; 431)

  "Developed from a symposium sponsored by the Division of
Chemical Information at the 198th Meeting of the American
Chemical Society, Miami Beach, Florida, September 10–15, 1989."

  Includes bibliographical references.
  ISBN 0–8412–1814–5

  1. Environmental monitoring—Data processing—Congresses.
2. Expert systems (Computer science)—Congresses.
3. Microcomputers—Congresses.

  I. Hushon, Judith M. 1945–   . II. American Chemical
Society. Division of Chemical Information. III. American
Chemical Society. Meeting (198th: 1989: Miami Beach, Fla.)
IV. Series.

TD193.E96   1990
363.7′ 063′0285633—dc20   90–37399
                                                   CIP

The paper used in this publication meets the minimum requirements of American National Standard for Information Sciences—Permanence of Paper for Printed Library Materials, ANSI Z39.48–1984.

PRINTED IN THE UNITED STATES OF AMERICA

ACS Symposium Series

M. Joan Comstock, *Series Editor*

1990 ACS Books Advisory Board

Foreword

The ACS SYMPOSIUM SERIES was founded in 1974 to provide a medium for publishing symposia quickly in book form. The format of the Series parallels that of the continuing ADVANCES IN CHEMISTRY SERIES except that, in order to save time, the papers are not typeset but are reproduced as they are submitted by the authors in camera-ready form. Papers are reviewed under the supervision of the Editors with the assistance of the Series Advisory Board and are selected to maintain the integrity of the symposia; however, verbatim reproductions of previously published papers are not accepted. Both reviews and reports of research are acceptable, because symposia may embrace both types of presentation.

Contents

Preface

ENVIRONMENTAL EXPERT SYSTEMS PROVIDE a new tool in the workbox of persons responsible for solving environmental problems. These systems already receive broad acceptance in a number of areas and are becoming the methodology of choice for such operations as management of sewage treatment plants and costing of hazardous waste site cleanup.

Much credit for encouraging the development of expert systems goes to Dan Greathouse, Lew Rossman, and John Convery of the Environmental Protection Agency (EPA) in Cincinnati and to Darwin Wright of EPA's headquarters in Washington, DC. They had the foresight to recognize the applicability of expert systems to environmental problem solving and the means to encourage development of systems to meet some of EPA's needs. From June 16 to 18, 1987, they organized in Cincinnati a meeting of people interested in environmental expert systems, providing the first chance for many of us to share our accomplishments and to experience one another's systems.

The group immediately recognized the utility of that initial meeting as a means of defining EPA standards for expert system development, of learning about applicable tools and methodologies, and of learning what is being done in the field. Since that time, the group has met at least annually, usually in connection with the Hazardous Materials Control Resources, Inc. (HMCRI), meetings in Washington, DC, in late November of each year.

During 1988, a number of us discussed the need to publish our approaches and methodologies for building and testing environmental expert systems as well as information on our existing systems. We felt it was especially important that this information be made available to the academic community since many students interested in selecting thesis topics related to environmental expert systems were having trouble determining the status of developments in the field.

The symposium on which this book was based was the first opportunity of the group not only to present orally their systems and research areas but also to prepare papers for publication. This book contains those papers as chapters.

The first chapter presents an overview of the state of the art of environmental expert system development. The chapter describes the system platforms, languages, and trends in system development. It is followed by several general chapters focusing on life cycle management as the process by which expert systems should be developed, the procedures for knowledge acquisition, system verification and validation, and neural network models as a specialized area of intelligent system development and how they can be applied to help address environmental problems.

The next group of chapters is focused on application areas for expert systems with such fields as sampling and analysis, predicting aquatic toxicity, assistance for operation of publicly owned treatment works, models for supporting the defining of problems and managing hazardous waste site operations, and site ranking models. Each system description includes emphasis on problems encountered during development and how they were solved. The final chapter defines the needs identified within EPA for expert systems and provides a vision of some areas of future environmental expert system development.

Together, these chapters present an important summary of the majority of the current work in environmental expert system development and represent most of the efforts actually being commercialized. This is a rapidly evolving field, but one with significant paybacks in terms of providing environmental problem solutions to end users faster and with less hassle and expense.

JUDITH M. HUSHON
Roy F. Weston, Inc.
955 L'Enfant Plaza, S.W., Sixth Floor
Washington, DC 20024

April 26, 1990

Chapter 1

Overview of Environmental Expert Systems

Judith M. Hushon

Roy F. Weston, Inc., 955 L'Enfant Plaza, S.W., Sixth Floor,
Washington, DC 20024

While expert systems technology has now existed for more
than 20 years, environmental expert systems are only
about five years old. Nonetheless, the development has
been rapid with over 68 systems in existence today. All
of the early systems and the bulk of the current systems
are PC-based, but as the limitations of the delivery
capability are reached, more and more systems are moving
toward larger delivery environments such as minicomputers
and dedicated workstations. Development is occurring
both using Artificial Intelligence languages such as
Prolog and LISP as well as expert system "shells." The
problems being tackled are also expanding. Whereas a
number of the early systems took on very limited areas of
expertise, such as the operation of a sewage treatment
plant, the systems are now moving out to tackle siting
problems and recommendation of complex remedial technology combinations. What is even more important is that
expert systems are becoming an accepted vehicle for
offering advice for solving environmental problems. Over
the next few years more complex systems will be developed
that share databases and tackle multiple related environmental problems.

Expert System development began in the late 1960s, but the first
systems were not completed and demonstrated until the early 1970s.
These systems generally sought to solve problems in narrowly defined
areas that were well understood by a few experts. The earliest and
most published system is MYCIN that was developed at Stanford
University to help diagnose and identify drug therapies for treating
pulmonary bacterial infections.(1) Another early system was PROSPEC-
TOR developed by SRI to assist field geologists in identifying
promising geographies for preliminary drilling for mineral deposits.
(2) Other important systems included XCON developed by Digital
Equipment Corporation to facilitate computer system configuration (3)

0097–6156/90/0431–0001$07.00/0
© 1990 American Chemical Society

and Delta/Cats-1 developed by General Electric and used to help diagnose faults in diesel-electric locomotives.(4)

In 1987, David Waterman published a book listing over 181 systems in the fields listed in Table I.(5) The environmental areas are notably missing with the possible exception of meteorology; and weather models had been under development for years. There are probably two reasons for the relatively slow emergence of expert systems in the environmental area. The first is that the science for dealing with environmental problems is not well understood and there are few absolutely agreed upon methods. This is in contrast to medicine where a medical textbook is considered the "Bible" and even the experts follow its advice exactly. The second reason is that few environmental problems can be solved by a single expert. There is often a need to involve environmental, civil, and chemical engineers, environmental chemists, and toxicologists to identify an optimal problem solution. The problems in trying to incorporate the knowledge of these multiple experts into a system are significant.

Nevertheless, environmental expert systems have begun to appear. In February of 1987, Hushon identified 21 environmental expert systems.(6) By December of that year, the number had risen to 51.(7) A current count puts the number of systems at about 69 A graph showing this growth is shown in Figure 1; by 1990, there will be close to 80 systems.

Development is occurring in Europe and Canada as well as the U.S. A recent review article by Page details development of 21 systems, mostly in Canada and West Germany.(8) Several other review articles have recently appeared (9, 10, 11).

Definition of Expert Systems

Expert systems are generally considered to be a branch of artificial intelligence; with their knowledge base, these systems can function as "experts" to make higher-level decisions based on varying performance levels. Expert systems have been defined as "man and machine systems with specialized problem-solving expertise;" each relies on a database of knowledge about a particular subject area, an understanding of the problems addressed within that subject area, and skill at solving these problems.

Expert systems are distinguished from traditional data processing systems in several ways:

o   they perform difficult tasks at expert performance levels.

o   they emphasize problem solving strategies.

o   they employ a certain amount of self knowledge to evaluate their own inference mechanisms and justify their conclusions.

o   they can deal with both symbolic and numeric logic.

o   they provide for the consideration of incomplete or uncertain data sets.

o   they provide justifications for their conclusions.

o   they also follow the human consultation paradigm.

Expert systems can vary in the type of logic that they use in
solving the problems. Two predominate approaches are known as forward
and backward chaining. If the search for a solution is started from
a set of conditions or basic ideas and moves toward some conclusion,
this is called forward chaining. In forward chaining, one starts with
known data and infers conclusions to reach an ultimate goal. The
logic works by taking as given the IF part of an IF...THEN rule and
inferring that the THEN parts are true. It then looks for rules in
which the THEN condition of the first rule is an IF condition in
another rule; several levels of inference may be involved. It should
be pointed out that forward chaining can be time consuming and can
lead to multiple conclusions.
Backward chaining attempts to determine if a stated goal rule is
satisfied by starting with the THEN clauses and backing up to the IF
clauses of the rule to see if they are fulfilled and so on until a
question is asked or a previously stored result is found.
These differences must be considered in choosing an approach for
developing a new expert system and in selecting an expert system
development tool. Forward chaining is preferred for identifying
options while backward chaining is preferred for identifying whether
specific options are viable.

Construction of an Expert System

The stages in construction of an expert system have been defined as:
system design, system development, formal evaluation of performance,
formal evaluation of acceptance, extended use in a prototype environment, development of maintenance plans, and system release.(12)
It is the job of the knowledge engineer to query the experts to
identify what information they employ to solve the problems being
modeled and how they combine this information to reach a conclusion.
It is then his/her job to incorporate this knowledge into the expert
system by writing the necessary software. The system may either
contain or must know how to access the databases of information it
requires. In addition, the knowledge base consists of a set of
IF...THEN rules or other knowledge representation methods such as
frames [Knowledge representation method that associates features with
nodes representing concepts or objects. The features are described
in terms of attributes and/or objects. All members of a common frame
have a similar set of attributes.] or semantic nets [Knowledge
representation method consisting of a network of nodes standing for
concepts of objects connected by arcs describing the relations between
nodes.] that describe how the expert combines the various decision
making parameters. The inference engine is the software that provides
the mechanism for interpreting the commands and accessing the
knowledge base to solve the problem.

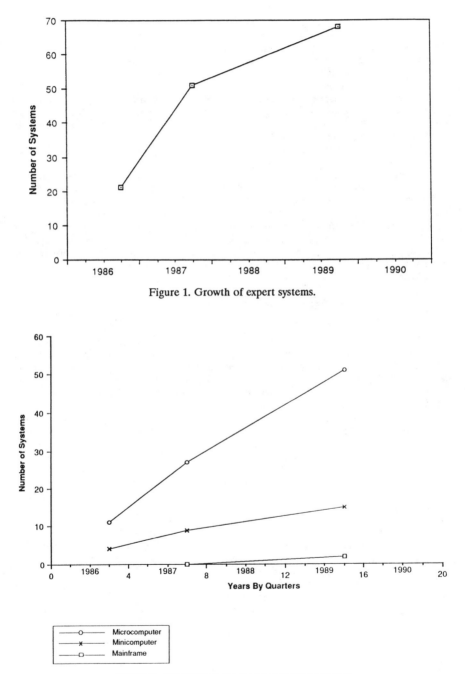

Figure 1. Growth of expert systems.

Figure 2. Environmental expert systems hardware.

Problems that Lend Themselves to Expert System Solutions

In identifying problems that lend themselves to solution using expert systems, it is often useful to try to determine whether the following characteristics apply:

o   Situations occur often

o   Situations are complex

o   Knowledge of experts required (higher reasoning)

o   Uncertainty involved

o   Situation is dynamic

o   Need to achieve consistency of response.

Expert Systems are included in a class of systems known as knowledge-based systems and full implementations may involve including more than one type of system. Other knowledge-based system tools include hypertext, which provides underlying links and is ideal for providing occasional access to help screens, diagrams or databases, and neural nets which use other types of logic to solve problems (in essence, the system develops the relationships among the variables and then uses these relationships to decide how best to handle new cases). Typical disciplines in which expert systems have been applied are shown in Table I.(5)

Table I. Application Areas for Expert Systems

Agriculture	Manufacturing
Chemistry	Mathematics
Computer Systems	Medicine
Electronics	Meteorology
Engineering	Military Science
Geology	Physics
Information Management	Process Control
Law	Space Technology

SOURCE: Reprinted with permission from ref. 5. Copyright 1986 Addison–Wesley Publishing Company, Inc.

Within a given discipline, certain categories of systems tend to arise sooner and others later. For example, systems for design and plan development arise early while training systems tend to be later to develop.

Survey of Environmental Expert Systems

While it is possible to group environmental expert systems under the
traditional areas of development shown in Table II, an expansion of
the categories of system development was used to better reflect the
environmental areas of application. The 69 systems identified to date
are presented in Table III. This table provides a short description
of each system and information on the software and hardware environment, an indication of who developed the system and where, and a
citation if the system has been described in the open literature.

Table II. Traditional Areas of Expert System Development

Interpretation	Planning
Prediction	Diagnosis
Repair	Training
Design	Monitoring
Control	Debugging

The earliest environmental expert systems were in the areas of
diagnosis and planning for fairly narrow applications. As development
progressed, there were more systems in the interpretation area and in
design. These systems generally require a broader knowledge base and
are more detailed, so it is not surprising that they were slower to
develop. Examples of these types of systems include those for
emergency response and those for remedial alternative selection.
Training systems tend to develop later still, often in the same areas
as the expert decision systems; only one expert system training aid
has been identified, this one in the water resources area.

Development Environment

The development environment for expert systems includes both hardware
and software. Initially, most of the expert systems were developed
on microcomputers. However, as shown in Figure 2, there is a steady
increase in the number of minicomputer or workstation based systems.
This is due to a variety of factors, the most common of which is that
the PC systems run out of computer "space" before they can solve a
complex problem due to the size of the code and other operating
requirements. The IBM compatible PC-AT is still the most common
development platform because it is a very widely distributed system
and provides the broadest user base. However, the limit of 640K of
random access memory is causing programmers to undertake ingenious
solutions to fit their code into this space.
    This problem has led one major software vendor to develop a code
that can be developed on a minicomputer, but delivered in a portable
PC environment.(44) This allows the developer access to the extra

Table III. Survey of Environmental Expert Systems

SYSTEM NAME	FUNCTIONALITY/STATUS	SOFTWARE	HARDWARE	DEVELOPER	CITATION
Selection of Service Providers					
1. LABSYS	Selection of analytical labs for environmental samples (in use)	DM	Mini	J. Hushon/WESTON	6
2. TSDSYS	Selection of treatment/recycle facilities for wastes (in use)	DM	Mini	J. Hushon/WESTON P. Hawkins/EPA/ERD	6
Selection of Remedial Action Technology					
3. Toxic Waste Advisor	Cleanup technology selection for solvents and hydrocarbons (prototype)	PC-Plus	Micro	J. Keenan/U. of PA	13
4. RPI Site Assessment System	Helps to characterize a waste site using the MITRE HRS for permeability and groundwater flow (prototype)	OPS5	Mini	K. Law/RPI	14
5. GEOTOX*	Computes a site hazard ranking based on key site parameters (prototype)	PROLOG	Micro	G. Mikroudis/WESTON H. Fang/Lehigh Univ.	15
6. Technology Screening System*	Works as a front-end to CORA to screen remedial sites. Based on engineering, scientific considerations, waste type, etc. (prototype)	Insight 2+	Micro	J. Crenca/CH2M-Hill K. Biggs/EPA/OERR L. Rossman/EPA/RREL	11

Continued on next page

Table III. *Continued*

SYSTEM NAME	FUNCTIONALITY/STATUS	SOFTWARE	HARDWARE	DEVELOPER	CITATION
7. TSAR	Helps select remedial technologies during RI phase of cleanup and suggests additional required input data for final selection in FS phase (prototype)	KEE	Mini	N. Pandit/WESTON R. Cibulskis/EPA/REAC	
8. DEMOTOX	Estimates groundwater pollution from leached wastes (prototype)	M.1	Micro	P. Ludvigsen/ERM	16
9. ARARS Screening	System to aid in determining requirements that drive the selection of remedial alternatives for Superfund sites (planned).		Micro	N. Pandit/WESTON D. Greathouse/EPA/RREL	9
10. XUMA	Uses chemical data to evaluate hazards and to identify disposal alternatives (in use)	ART	Mini	R. Weidemann/Nuclear Research Center Karlsruhl/FRG	17
11. XSAL	Identification, analysis, and evaluation of waste sites to provide assistance on remediation (prototype)	PROLOG C	Mini	H. Groh/Technical College Saarbrucken, FRG	18
12. ALEXIS	Used for identification, inventory, evaluation and monitoring of contaminated istes and remediation (prototype)	IBM ES Shell Env.	Main IBM/370	H. Franzen/German Assn. for Plant Survey	19

Table III. *Continued*

SYSTEM NAME	FUNCTIONALITY/STATUS	SOFTWARE	HARDWARE	DEVELOPER	CITATION
Plan Generator					
13. Work Assignment/Work Plan Generator	Uses inputs about a waste site to plan activities and generate draft workplan (in use)	OPS5	Mini	J. Schmuller/CDM	20
14. HASP*	This health and safety plan generator will assist in preparation of health and safety plans for Superfund sites. It relies on activities, chemicals and quantities. (prototype)	Knowledge-Pro	Micro	J. Hushon/WESTON R. Turpin/EPA/REAC	
15. CES	This closure evaluation system consists of three systems to aid in review of closure plans regarding vegetation cover, final cover, and leachate collection system (prototype).	PC-Plus Knowledge-Pro	Micro	D. Greathouse/EPA/RREL	
16. Computerized System for Community Planning	Uses inputs on local industries and the materials present to rank potential community hazards for CEPP (prototype)	dBase III	Micro	J. Bare/EPA/ESRL	6
17. Smart Methods Index*	Supports natural language query of database to identify analytical methods for particular waste site pollutants (prototype)	PROLOG	Micro	R. Olivero/Lockheed D. Bottrell/EPA/EMSL	
18. CORA*	Helps estimate costs for cleanup of a waste site (in use).	Insight 2+	Micro	K. Biggs/EPA/OERR J. Crenca/CH2M/Hill	11

Continued on next page

Table III. *Continued*

SYSTEM NAME	FUNCTIONALITY/STATUS	SOFTWARE	HARDWARE	DEVELOPER	CITATION
19. SCEES	Uses HRS data to identify work required to cleanup a waste site in terms of cost and schedule (in use).	NEXPERT	Micro	J. Schmuller/CDM	10
20. Super Disc	System to aid in review of claims by contractors for contract modification due to differing site conditions (planned)	Knowledge-Pro	Micro	L. Bennett/CDM D. Greathouse/EPA/RREL	
21. IQAP*	This system will provide assistance to develop QA plans for Superfund sites (prototype)	Hypertext	Micro	N. Pandit/WESTON R. Cibulskis/EPA/REAC	
Facility Siting					
22. Hazardous Waste Facility Siting	Helps determine siting in environments near wetlands (prototype)	M.1	Micro	V. Lambou/EPA/EMSL	11
Problem Diagnosis					
23. Activated Sludge Diagnosis	Uses instrument and lab test results to facilitate the operation of an activated sludge wastewater treatment facility (prototype)	?	Micro	D.M. Johnston/University of Washington	21
24. Waste Incineration	Diagnosis malfunctions in hazardous waste incinerators (prototype)	M.1	Micro	Y.W. Huang/University of Houston	22

Table III. *Continued*

SYSTEM NAME	FUNCTIONALITY/STATUS	SOFTWARE	HARDWARE	DEVELOPER	CITATION
25. Incineration Process Control*	System to control conditions required for destruction of hazardous wastes in a model incinerator (prototype)	OPS5	Micro	C. Subramanian/WESTON	
26. POTW Expert*	Expert system to diagnose performance limiting factors at publicly owned treatment works (prototype)	ALEX Smalltalk/V	Micro	L. Berkman/Eastern Research Group	
27. Activated Sludge Advisor*	To assist waste water treatment facility operators to manage an activated sludge system (prototype)	Knowledge-Pro	Micro	J. Schmuller/CDM	
28. Water Treatment Plant	Assists operators in assessing water quality and quantity; suggests required maintenance (prototype)	KEE	Mini	S. Nix/Syracuse University	11
29. Expert System for Diagnosis of Wastewater Treatment Plants	Assists operators to diagnose faults in conventional activated sludge treatment plant (development)	PC-Plus	Micro	G.G. Patry/McMaster Univ.	23
30. Anaerobic Digestion System	Problem diagnosis due to poor stability (prototype)	M.1	Micro	M. Barnett/Rice University	11
31. Sludge Cadet	Diagnose problems in trickling filter systems and suggest remedies (prototype)	KEE	Micro	C. Perman/Stanford Univ.	11

Continued on next page

Table III. *Continued*

SYSTEM NAME	FUNCTIONALITY/STATUS	SOFTWARE	HARDWARE	DEVELOPER	CITATION
32. Computer Aided Data Review and Evaluation (CADRE)*	Reviews Laboratory data on volatile and semi-volatile organics and pesticides as part of data evaluation (in use)	PROLOG	Micro	R. Olivero/Lockheed D. Bottrell/EPA/EMSL	
33. Environmental Sampling Expert System (ESES)*	System recommends sampling procedure, locations, numbers of samples and handling procedures for soil sampling for metals (prototype)	Knowledge-Pro	Micro	R. Olivero/Lockheed D. Bottrell/EPA/EMSL	
34. Leaking Underground Storage Tank System	Provides remedial action for controlling tank leaks	Gespe	Micro	D. Marks/MIT	11
35. Dike Maintenance System	Advises on repairs and Maintenance of dikes	SAGE	Micro	SDS/Rykwaterstaat Aviesdients	24
36. Environmental Assessment System (EASY)	Assesses environmental impact and implications of manufacturing processes to achieve waste reduction (in use)	KEE LISP	Micro	E. Venkataramani/Merck & Co., Inc.	25
Permit Assistants					
37. SEPIC	Issues permits for onsite private sewage disposal systems (in use)	Rulemaster	Micro	W.J. Hadden/Intelligent Advisors, Inc.	26
38. Permit Writer's Assistant	Assists EPA in issuing water permits for several industry sectors (in use)	KES	Micro	C. Spooner/EPA/OWP	27

Table III. *Continued*

SYSTEM NAME	FUNCTIONALITY/STATUS	SOFTWARE	HARDWARE	DEVELOPER	CITATION
39. Expert System for Assistance in Handling Environmental Regulations	Used to help identify applicable regulations and documents related to permitting (in use)	LISP	Mini	M. Halker/ F. Bubeck/Siemens GmbH.	28
40. WAPRA	Assists EPA to review waste analysis plans that are part of part B permit applications. Screens for potential chemical incompatibilities (in use).	PROLOG	Micro	D. Greathouse/EPA/RREL	6
Model Front-Ends					
41. QUAL2E Advisor	Helps user determine input parameters for QUAL2E surface water quality simulation model (prototype)	M.1	Micro	T. Barnwell/EPA/ERL	29
42. INHEC-1	Helps user select input parameters for the HEC-1 groundwater model (in use)	KES	Micro	N. Pandit/WESTON	
43. Storm Water Management Model Calibrator	Help users calibrate the EPA SWM model (in development)	KES	Micro	J. Delleur/Purdue University	11
44. EXSRM	Used to estimate initial input parameters for snow runoff model (prototype)	ART	Mini	E.T. Engman	30
45. Flood Advisor	Helps select an appropriate flood estimation model (prototype)	C	Mini	D. Fayegh	31

Continued on next page

Table III. *Continued*

SYSTEM NAME	FUNCTIONALITY/STATUS	SOFTWARE	HARDWARE	DEVELOPER	CITATION
46. Expert Rokey	Assists hydrogeologists to estimate subsurface distribution of chemicals discharged from underground sources (prototype)	FORTRAN C	Micro	Simco Groundwater Research Ltd.	32
47. RAISON	Interface to regional acid rain model to determine water quality.	C	Micro	D. Lam/Univ. of Guelph	33
48. Mixing Zone Analyzer	Selects appropriate water quality model for use in mixing zone	M.1	Micro	G. Jirka/Cornell University	11
49. Groundwater Flow Analyzer	Calibrates input parameters to flow models.	PROLOG	Micro	A. Frank/Univ. of Maine	11
Engineering Tools					
50. FLEX	Assists in selection of flexible membrane liners for landfills and surface impoundments (prototype)	PROLOG	Micro	L. Rossman/EPA/RREL	34
51. MUMS	Assists dam gate operators to control water flow from reservoir systems (in use)	KES	Mini	N. Pandit/WESTON	6
52. REZES	Advisor to support reservoir management and operation (prototype)	PROLOG FORTRAN	Micro	University of Manitoba	35
53. FIESTA	Helps evaluate sludge test results (demonstration)	EXSYS	Micro	N. Pandit/WESTON	

Table III. *Continued*

SYSTEM NAME	FUNCTIONALITY/STATUS	SOFTWARE	HARDWARE	DEVELOPER	CITATION
54. RESREC	Helps evaluate resource recovery options based on waste characteristics and costs. (in use)	VP-Expert	Micro	N. Pandit/WESTON	
Risk Assessment Tools					
55. Risk* Assistant*	Helps assess human health risks posed by hazardous waste (prototype)	C	Micro	J. Young/Hampshire Res. Inst J. Segna/EPA/OHEE	
56. SCREENER	Used to evaluate/predict impacts at airports for Environmental Assessments (in use)	PROLOG C	Micro	R.R. Everitt G.D. Sutherland/ESSA Ltd.	36
57. Aid for Evaluating the Redevelopment of Industrial Sites	Risk assessment model that uses environmental and toxicological data to estimate exposure at redevelopment sites (in use)	Level 5	Micro	B. Ibbotson/SENES Consult.	37
58. Expert System for Identifying Biological Species	Used to identify endangered species. Linked to optical disk (prototype)	OPS5	Main (VAX)	G. Hille/Univ. of Hamburg	38
Emergency Response Tools					
59. FRES	Assists first responders to chemical emergencies by identifying hazards and suggesting response methods (prototype)	KES	Micro	J. Hushon/WESTON	39

Continued on next page

Table III. *Continued*

SYSTEM NAME	FUNCTIONALITY/STATUS	SOFTWARE	HARDWARE	DEVELOPER	CITATION
60. HERMES	Emergency response support system to assist in chemical incidents (prototype)	ART	Mini	E. Chang/Alberta Research Council	31 40
61. CORKES	Community right to know system combines inventory with emergency response assistance (prototype)	KES	Micro	J. Hushon/WESTON	6
62. Expert System Dispatch for Forest-Fire Control Resources	Dispatches forest fire control equipment and crews based on incomplete data (in use)	PROLOG FORTRAN	Mini	A. Gray/Canadian Forestry Service	41
63. FEUEX	Advice on transport of hazardous materials and fire-fighting (prototype)	PROLOG LISP	Mini	F. Belli/Technical College Bremerhaven	42
Toxicity Prediction					
64. Fish Toxicity Prediction*	System uses chemical structure to predict aquatic toxicity (in use)	LISP	Micro	J. Hickey/DOI/NFRC-GL	
Site Ranking					
65. Defense Priority Model (DPM)*	System to rank DOD hazardous waste sites for cleanup based on their potential threats to health and ecology (in use).	PROLOG	Micro	J. Hushon/WESTON A. Kaminski/USAF	
66. Multi Media Environmental Pollutant Assessment System (MEPAS)*	System to assess environmental concerns i.e., potential human impacts from DOE site. (in use)	FORTRAN C	Micro	J. Droppo/Battelle PNL	

Table III *Continued*

SYSTEM NAME	FUNCTIONALITY/STATUS	SOFTWARE	HARDWARE	DEVELOPER	CITATION
Teaching Systems					
67. Water Resources Lab Aid	Tutors students in use and calibration of a series of water models (in use).	EXSYS	Micro	R. Carlson/Univ. of Alaska	11
Regulatory Evaluation					
68. Expert System for Hazardous Waste Regulations	System to help identify appropriate waste regulations (development)	Rulemaster	Micro	P.A. Barrow/Univ. of Alberta	43
69. Smart reg (r)	Provides assistance in interpreting Underground Storage Tank regulations (in use).	Black Magic Basic	Micro	M. Stunder/GEOMET	

*Detailed paper in this volume.

tools and compiling space afforded by the workstation, while making it possible to distribute the system to end users with less powerful hardware.

Software comprises the other major component of the development environment. System developers have used two routes - shells and languages. Shells are computer languages developed to facilitate the development of expert systems. They allow the user to write programs in English-like grammars and provide functionality for screen management, calling programs, computing uncertainty, and the problem solution strategy. They exact a price for this functionality, however. They remove a number of design options from the developer and force expert system development within their rigidly defined environments; they also run more slowly due to the extra translation step. Shells are particularly useful for prototyping a system to see if the problem can be solved using expert systems technology. The code can be rewritten later if the concept can be proven. PC shells tend to be much more restrictive than the shells on the workstations.

Table IV shows the most commonly used shells for environmental expert system development, and the number of systems developed in each.

The alternative to using shells is to develop the expert system directly in a high level computer language. This may be either an Artificial Intelligence (AI) language such as Prolog or LISP or a standard language such as C or FORTRAN. In fact, most expert systems shells are built on top of these languages. Table V shows the languages to develop environmental expert systems. If a system used more than one language, this was noted in Table III and it is included twice here, once for each language.

Another advantage of developing systems directly in languages is that the code can be compiled and distributed, protecting the source code. It is also an advantage to be able to be able to distribute the compiled code directly and not have to worry about the user's need to purchase "run time software" which is required to make many of the expert system shells useful. Some shell software vendors are moving toward charging higher prices for the development code and providing free run time code and others are providing inexpensive licenses for run time code that require only a one time purchase.

Legal Issues Associated with Expert Systems

The developers of expert systems are concerned about the whole issue of liability. According to the Brookings Institution, the number of product liability lawsuits has increased eightfold from 1974 to 1986 when 13,595 such cases were filed.(44) However, in the expert systems area, the first suit has not yet been filed. Some system developers are attempting to limit their responsibility by including disclaimers, but lawyers say these offer little refuge because buyers rarely return license agreement cards. Lawyers have suggested that systems which leave the final decision up to the user will have reduced liability. It is also likely to depend upon whether there is a "bug" in the software; no software company has yet lost a lawsuit brought over a bug though there have been several out of court settlements. One final claim used by developers is that expert systems represent an inexact science.

Table IV. Expert System Shells

Shell	Number
Minicomputer/Workstation Shells	
KEE	4
ART	3
IBM Expert System Shell	1
Microcomputer Shells	
M.1	6
KES	6
Knowledge-Pro	5
OPS5	4
PC-Plus	3
Insight 2+	2
Exsys	2
Rulemaster	2
Level 5	1
Sage	1
Hypertext	1
Nexpert	1
Gespe	1
ALEX	1
Black Magic	1

Table V. Languages Used to Develop Environmental Expert Systems

Language	Number	Percent
Prolog	12	39
C	7	23
FORTRAN	4	13
LISP	3	9
DM	2	6
Smalltalk/V	1	3
Basic	1	3
dBase III	1	3

There are also efforts underway that are focused on reducing the total liability to software developers by avoiding excessive awards and introducing a no-fault claims system which may be of future benefit.(45)

These concerns have slowed private systems development and have led universities to examine their potential liability. As a result, many are content with only developing prototypes which never become production systems. Another result is that much of the existing development has had government cooperation in an attempt to limit liability by the private developer.

Future Trends

Acceptance of expert systems by the environmental community is increasing which supports the view that these types of systems are here to stay and will play an increasingly important role as the demand for smarter systems grows. While the initially developed systems tackled small and well understood problems, there is a trend toward trying to solve more complex problems in areas where there is a higher degree of uncertainty. In these areas, expert systems are being used to provide "gut reactions" to problem solutions just as we ask experts to do. These systems will also be used increasingly to deal with incomplete data sets.

There is already a trend toward more complex systems with larger databases. The databases may be located on the computer with the expert system or they may be remotely accessed and the required data downloaded by a small subroutine called by the system.

The more complex systems demand larger hardware to function maximally. The standard IBM compatible PC-AT with 640K of RAM is no longer sufficiently large to handle the complex applications being contemplated. This means that the more complex systems will have to be developed and used on larger machines which will limit their availability. This is perhaps the most critical developmental criterion affecting the future of expert systems application to environmental proglems.

Not surprisingly, the initial development of environmental expert systems took place at universities across the country. While significant development has continued at the university level, there is increasing activity among private consulting firms to develop products, often with government funding. This is an important step, for it means that systems are being employed to solve the problems for which they were developed. This increase in government funding has also brought with it a standardization of the system development and evaluation methodology.

The number of environmental expert systems can be expected to increase rapidly for at least several years since there are many problems for which expert systems can provide superior solutions to those available from traditional computer programs. And as the solutions become better, the built in knowledge of the systems will become increasingly transparent to the user.

References

1. Shortliffe, E. H.; Buchanan, B.G.; Fiegenbaum, E. A. Proceedings of the IEEE, Vol 67, 1979, pp. 1207-1224.

2. Duda, R. O.; Gaschig, J. G.; Hart, P. E. In Expert Systems in the Micro-electronic Age, Mitchie, D., Ed.; Edinburgh University Press: Edinburgh, 1979, pp. 135-137.

3. McDermott, J. Artificial Intelligence, No. 19, 1982, p. 29.

4. Artificial Intelligence Report, Vol 1, No. 1, 1984, pp. 7-8.

5. Waterman, D. A Guide to Expert Systems, Addison Wesley: Reading, MA, 1986.

6. Hushon, J. M. Environmental Science and Technology, Vol 21, No. 9, 1987, pp. 838-841.

7. Hushon, J. M. AIChE paper.

8. Page, B. Environmental Computing, 1989 (in press).

9. Greathouse, D. In CRC Critical Reviews in Environmental Control, 1990 (in press).

10. Ortolano, L.; Steinemann, A. C. Journal of Computing in Civil Engineering, 1987, pp. 298-302.

11. Rossman, L. M. In Expert Systems for Civil Engineers: Technology and Applications, Maher, M. L., Ed; 1987, pp. 117-118.

12. Yaghmai, N. S.; Maxin, J. A. Journal of the American Society for Information Science, September 1984, p. 297.

13. Keenan, J. Design Specifications-Descriptions of Toxic Waste Advisor (TWA) Expert System; University of Pennsylvania, internal report, 1986.

14. Law, K. H.; Zimmie, T. J.; Chapman, D. R. In Expert Systems in Civil Engineering; Kostem, C. N.; Maher, M. L., Eds.; American Society of Chemical Engineers: New York, 1986; pp. 159-173.

15. Mikroudis, G. K.; Fang, H. Y. In Proceedings, 1st International Symposium on Environmental Geotechnology; Lehigh University: Bethlehem, PA, April 1986; pp. 223-232.

16. Ludvigsen, P. J.; Sims, R. C.; Grenney, W. J. In Proceedings of ASCE Fourth Conference on Computing in Civil Engineering; American Society of Chemical Engineers: New York, October 1986; pp. 687-698.

17. Weidemann, R.; Geiger, W.; Eitel, W. In Informatikanwendungen im Umweltbereich; Karlsruhe Symposium Proceedings, Volume 2; Jaeschke, A.; Page, B., Eds.; Springer Verlag; 1988, pp. 116-126.

18. Groh, H.; Ruttler, R. In Informatikanwendungen im Umweltbereich; Hamburg Symposium Proceedings; Valk, R., Ed.; Springer Verlag; 1988.

19. Franzen, H. In GI-Fachausschuss 4.6, No. 8; November 1988.

20. Paquette, J. S.; Woodson, L.; Bissex, D. A. In Proceedings, Superfund, '86; Hazardous Materials Control Research Institute: Rockville, MD, 1986; pp 208-212.

21. Johnston, D.M. In Proceedings, Computer Applications in Water Resources, American Society of Chemical Engineers: New York, 1985; pp. 601-606.

22. Huang, Y. W. In Expert Systems in Civil Engineering; Kostem, C. N.; Maher, M. L., Eds.; American Society of Chemical Engineers: New York, 1986; pp. 145-158.

23. Page, B. Proceedings, Envirosoft 88 - 2nd International Conference, Greece, Zanetti, P., Ed.; Springer Verlag, 1988; pp. 597-608.

24. CRC Systems, Inc.; CDM Federal Programs Corp. Proceedings of Workshop on Expert and Automated Systems in Hazardous Waste Management, Cincinnati, 1987.

25. Venkataramani, E.S.; House, M.J.; Bacher, S. "Implementation of an Expert System Based Environmental Assessment System (EASY);" paper presented at AIChE meeting. August, 1989.

26. Hadden, W. J., Jr; Hadden, S. G. In Proceedings, Expert Systems in Government Symposium; Karna, K. N., Ed.; MITRE Corp.: McLean, VA, 1985, pp. 558-566.

27. Spooner, C. S.   In Proceedings, Expert Systems in Government Symposium; Karna, K. N., Ed.; MITRE Corp.: McLean, VA, 1985, pp. 573-577.

28. Halker, M.; Bubeck, F. In Proceedings, 16th GI-Jahrestagung; Informatik-Fachberichte 127: Berlin, Heidelberg, New York, 1986, pp. 436-447.

29. Barnwell, T. O., Jr.; Brown, L. C.; Marek, W.   "Development of a Prototype Expert System for the Enhanced Stream Water Quality Model QUAL2E;" internal report, U.S. Environmental Protection Agency: Athens, GA, 1986.

30. Engman, E. T.; Rango, A.; Martinec, J.   In Peoceedings, Water Forum '86; American Society of Chemical Engineers: New York, 1986; pp. 174-180.

31. Fayegh, D.; Russell, S. O.   In Expert Systems in Civil Engineering; Kostem, C. N.; Maher, M. L., Eds.; American Society of Chemical Engineers: New York, 1986; pp. 174-181.

32. Proceedings of Workshop at Annual General Meeting of Canadian Prairie and Northern Section of the Air and Waste Management Association; Edmonton, Air and Waste Management Association, Ed., 1989.

33. Lam, D. C. L.; Fraser, A. S.   In Proceedings, Envirosoft 88 -2nd International Conference, Greece, Zanetti, P., Ed.; Springer Verlag, 1988; pp. 67-80.

34. Rossman, L. A.; Haxo, H. E., Jr.   In Proceedings, Environmental Engineering Specialty Conference; American Society of Chemical Engineers: New York, 1985; pp. 583-590.

35. Environment Canada, The Application of Artificial Intelligence (Expert Systems) in Environment Canada; Proceedings of a Workshop; Ottawa, June 1988.

36. Everitt, R. R.; Sutherland, G. D.   AI Applications in Natural Resource Management, Vol. 2, No. 4, 1988, pp. 55-56.

37. Ibbotson, B. G.; Powers, B. P.   Proceedings of Workshop at Annual General Meeting of Canadian Prairie and Northern Section of the Air and Waste Management Association; Edmonton, Air and Waste Management Association, Ed., 1989.

38. Hille, G.   In Informatik im Umweltschutz - Anwendungen und Perspektiven, Munich-Vienna, 1986.

39. Hushon, J. M.   FRES presented at 190th National Meeting of ACS, NYC April 1986.

40. Chang, E.; Clark D. Sidebottom, G.   In Proceedings, 5th Technical Seminar on Chemical Spills, Montreal, February 1988, pp. 323-335.

41. Gray, A.; Stokoe, P. Knowledge-based or Expert Systems and
     Decision Support Tools for Environmental Assessment and Manage-
     ment - Their Potential and Limitations, Federal Environmental
     Assessment Review Office, School for Resource and Environmental
     Studies, Dalhousie University: Halifax, Nova Scotia, June 1988.

42. Belli, F.; Bonin, H. In Informatikanwendungen im Umweltbereich;
     Hamburg Symposium Proceedings; Valk, R., Ed.; Springer Verlag;
     1988.

43. Barrow, P. A. In Proceedings, SCS Simulations Conference,
     Orlando, April 1988, pp. 343-348.

44. Warner, E. High Technology Business, Vol 8, No. 10, 1988, pp.
     32-36.

45. Ruby, D. PC Week, Vol 3, July 8, 1986, p. 49.

RECEIVED April 27, 1990

Chapter 2

Success Factors for Expert Systems

Dan Yurman

Information Management Staff, Office of Program Management and
Technology, U.S. Environmental Protection Agency, 401 M Street, S.W.,
(OS-110), Washington, DC 20460

EPA's hazardous waste program has initiated a 5-year, $5
million program in to build expert systems. The Information Staff developed guidance adapting life cycle management practice to the rapid prototyping cycles of expert
systems. The "Practice Paper" has become a landmark document
for Federal civilian agencies engaged in developing expert
systems. This paper reviews the objectives and success
factors for expert systems contained in the guidance.

Background

Working in cooperation with EPA's Risk Reduction Laboratory at
Cincinnati, OH, the hazardous waste program has committed to a 5-year,
$5 million program to build expert systems. The key issue is not so
much feasibility - this technology works - but rather that the systems
built are reliable. The emphasis in this paper is on building reliable
expert systems.

Testing expert systems addresses two concerns - 1st; did the right
system get built?; and 2nd; was the system built correctly? This means
testing not only the software, but also the expertise embedded in the
system. Equally important is whether the organization, or the client,
has the capacity, discipline, and tenacity to see the system through
to completion. A prototype is neither a decision to ship a system nor
is it even a decision to build a system. The development and testing
of expert systems cannot be isolated from the other management issues,
and, more importantly, the organizational climate in which the work
is taking place.

In Spring 1987 the Information Management Staff (IMS), Office of
Solid Waste and Emergency Response (OSWER) at US EPA reviewed the
procedures being used to test expert systems. What was found there
were inconsistent controls on software testing of expert systems.
Additionally, legal, technical, and moral issues were mixed in to form

Current address: Idaho National Engineering Laboratory, EG & G Idaho, Inc., P.O. Box 1625,
Idaho Falls, ID 83415

a question of whether the work should continue. It was also discovered that there were more than one kind of testing problem The first problem was premature access by field users to beta test systems. The second problem centered in independent developers who failed to understand the need for field validation of the "expertise" in their systems. One professor of computer science at a major university announced completion of an expert system for hazardous waste cleanup, but admitted that he had never spoken with nor shown his system to any EPA field office staff. These situation could cause EPA a lot of problems.

There were no caveats or warnings in either case that the users were dealing with test data, or "naked numbers," and betatest software. Some of the users of the beta versions of the expert systems turned out to be companies regulated by EPA. Third, the potential existed for any one of the users to capture the data and use it in a permit application proceeding or as part of the defense against an enforcement action. In the private sector, these risks would be equal to sending key chapters of your company's business plan to the competition.

In response to this situation OSWER sent a policy document to field offices explaining the risks of using beta systems or prototypes in field offices. It emphasized that a clear audit trail is required with all expert systems because judgmental processes are involved. Users will be more confident in the recommendations of an expert system if the logic train is well documented. Expert systems must undergo both validation and verification processes. Developers must test both the code and the expertise contained in expert systems. It is advisable that all expert systems be validated by objective, third party reviewers or other experts. Experts might not always agree with the expert system. In lieu of peer review, management must use their authority in accepting an expert system application. Regarding legal issues, it is advisable that expert systems developed by government agencies be developed so that they are able to withstand the scrutiny of their many "publics."

Life Cycle Management

The next step in the process of getting control of expert systems was tied to OSWER's internal guidance on life cycle management. This guidance is based on the Federal Information Processing Standards (FIPS) publications. In September 1988 OSWER completed a management guidance document which merges traditional life cycle management practices with the special requirements of knowledge engineering and expert system development methods, including rapid prototyping.

As the project officer for this task, I had the responsibility for the final document. After receiving the document from the contractors in September 1988, OSWER engaged in an extensive internal review and also arranged for third party review outside the agency. The final guidance document was released in November 1988.

The implementation of life cycle management is based on decisions a manager must make at each stage in the product development cycle. The decisions seem obvious when applied to conventional applications. Although expert systems offer great benefits, care must be taken in selecting appropriate applications and in planning and monitoring development. Major issues include the need to:

o Identify problem areas suitable for expert systems.

o Determine the feasibility of applying expert systems to a
   particular situation given the available data and expertise.

o Determine whether or not using expert systems would result in
   productivity gains.

o Estimate resource requirements to develop, implement, and maintain expert systems.

Purpose.

The purpose of the "Practice Paper" on expert systems is to convey
guidance on the design, development, and operational issues for expert
systems. The objective is to find new ways to use computer technology,
and to avoid mistakes. This paper describes the use of life cycle
management principles and practices for developing expert systems.
The emphasis is on objectives for each of eight major stages and the
success factors for meeting these objectives.
    It is important to define and contrast expert systems with classical
decision support systems, executive information systems, and conventional
software systems. First business needs are identified, then appropriate
technologies are applied, similar to conventional systems. An analysis
of the cost effectiveness of a proposed expert system is important.

Capabilities of Expert Systems.

The project manager, software developer, and the users must understand
how knowledge processing in expert systems differs from conventional
data processing. Expert systems are unique in their ability to process
knowledge, not just data. Knowledge processing differs from data
processing by the type of information, the techniques to analyze the
information, and in the form that the results of the knowledge processing
are presented to the user.
    Conventional systems limit the developer to data representation
using only numbers and text. They process data using complex algorithms
that complete a discrete number of steps to reach a predetermined
conclusion. Expert systems permit knowledge representation - the
encoding of human decision-making processes using symbolic terms or
symbols. Because expert systems process knowledge, they are often
referred to as knowledge-based systems.
    The ability to represent knowledge in symbolic terms expands the
range of analysis techniques that computers can apply to information
thus enabling a system to emulate some aspects of human performance.
 The expert system uses problem solving procedures such as pattern-
matching to reason about the symbolic terms.
    The combination of problem solving procedures that are built into
expert systems, together with the developer's ability to define problems
using symbolic terms, give expert systems the capability to store and
manipulate more complex relationships between individual pieces and
groups of information than can be accomplished with the processing
supported by conventional systems. Expert systems can be designed with
the ability to explain the "reasoning" used in reaching a recommendation
and to justify their approach to a problem, much as people do.

Scope of Development Issues.

Issues for development of expert systems can be divided into eight general categories, which parallel the phases of life cycle management.

The first issues are cross-cutting concerns including project management issues that are addressed in multiple phases of the system life cycle. These are the project management plan, reviews and quality assurance, project approvals, configuration management, data administration, methodologies and tools, cost-benefit analyses, and knowledge management. The developer must acquire the resources required throughout the expert system life cycle.

1. The Initiation Phase for an expert system covers the tasks involved in problem definition and in determining the need for an automated solution. It explores characteristics of a problem that suggest an expert system solution. However, the conclusions drawn from this phase are usually written independent of any particular technology.

2. The Concept Phase involves the identification of a feasible, timely, cost-effective solution to the problem. This phase is used to determine information needed in the use of one or more proof-of-concept prototypes to refine the solution, knowledge representation, and management techniques, control structures, and justification for the chosen approach to system development.

3. The Definition and Design Phase involves decisions to confirm the suitability of the System Concept and to determine detailed functional requirements. It covers all aspects of designing the system and selecting a development environment including knowledge base creation, migration to delivery environments, and user interfaces.

4. The Development Stage addresses decisions in developing and using knowledge acquisition methodologies, sources, and conflict resolution techniques. The system is built at the production level in this stage. Also, it defines the means of testing and validating the system.

5. The Implementation Stage is used to identify the strategies for distributing expert systems. It includes the issues of beta testing, user registration, satisfying hardware and operating system requirements, training, licensing, documentation, configuration management, and version control.

6. The Operation Phase focuses on the Production, Evaluation, and Archive stages of the life cycle. It covers maintenance, end-user support requirements, knowledge revalidation options and maintenance, ongoing training and documentation, and software updates.

These stages are now described in terms of their objectives and success factors.

Objectives of the Initiation Phase.

The primary objective of the Initiation Phase is to describe the problem in clear, technology-independent terms upon which all affected business units can agree. The second objective for the Initiation Phase is to determine whether staff or other resources will be devoted to defining and evaluating alternative ways to respond to the identified problem in the Concept Phase. Committing resources beyond the Concept Phase is premature at this point. The use of rapid prototyping as a tool to work through these issues is a feasible approach at this stage and in the Concept Phase.

Success Factors for the Initiation Phase.

There are seven factors that can impede success if not considered in the Initiation Phase. They define what expert systems can and cannot do.

First, the developer must insure that there is not a bias toward building an expert system. It is important to determine that the problem has not already been solved to the satisfaction of the client by other types of conventional programming such as modeling, decision support, databases, linear programming, or text retrieval.

Second, expert system development projects must be reasonably scoped from initiation stage. Failure to do so can lead to developing a solution to the wrong problem, tackling overly complex problems, or attempting to solve nebulous problems.

Third, the developer must determine that the problem is well suited to use of expert system technologies. If the problem is purely algorithmic or procedural in nature, then it can be addressed by conventional technologies more efficiently than by expert systems. If the type of problem requires symbolic reasoning, then the problem may be suitable for expert systems technology.

Fourth, an expert must be available, and, the problem must be capable of being solved by conventional means. Some systems may incorporate the knowledge of more than one expert, while others reflect the knowledge and strategies of a single individual. It makes no sense to attempt to use an expert system to solve a problem if the answers are unknown. The reason is that it will be impossible to validate the expert system because users will not know if the system is providing correct answers. If no true expert exists, then the problem may be too nebulous and ill-defined to be effectively addressed by an expert system.

Fifth, the problem must have recognized bounds. It cannot have an infinite set of solutions. The solutions that will be considered by the system must be determined in advance. When the expert system attempts to work with information near the periphery of the problem domain, it may yield unpredictable results. Relying on an expert system in such circumstances could lead to catastrophe if the system is being used to monitor chemical manufacturing processes.

Sixth, an expert must be able to solve the problem by conventional means in less than an average work week. The project manager should also consider the complexity of the tasks that are to be automated with an expert system. The tasks should neither be too difficult nor too trivial for a human expert. A task requiring more than a week to solve without computer support is probably too large to be built using rules. While this does not eliminate other approaches, if the problem can be parsed into a group of linked, smaller problems, it may be manageable. On the other hand, a task requiring only a few minutes to solve might be automated more efficiently using conventional technologies.

Seventh, the investment in an expert system application must produce a payoff either in terms of improvement productivity or a measurable profit. Since expert system development requires a significant investment in terms of people and money, the expected return on that investment must be well understood, along with the means of measuring the return.

Objectives of the Concept Phase.

The objectives for the Concept Phase include problem definition, requirements and feasibility. These objectives define the approach to solve the information processing problem. The first objective of the Concept Phase is to confirm the existence of the information processing or knowledge-intensive problem. The second objective is to identify high level requirements for a solution to the problem. These requirements should focus on the nature of the problem and the user's needs. The third objective is to determine the feasibility of an expert system solution to the problem. This requires a study of the applicability of expert systems to the project and the capabilities of other information technologies in comparison to the choice of an expert system.

Success Factors for the Concept Stage.

Several factors that affect an expert system's success should be considered in the Concept Phase. They include organizational and resource issues, the target users of the system, functional requirements, and knowledge representation and control structure.

The first major success factor is to effectively implement the products of the Initiation Phase. Proper utilization of the results of the Initiation Phase leads to good results in the areas of assembling a team, requirements definition and resource estimation, and clear management direction.

The key organizational issue is management commitment. Management commitment to an expert system project is an important reason for project success. Management commitment comes in many forms: resources -- including dollars, people, and equipment; continued involvement and supervision; and support in times of conflicting objectives. Management needs to be aware of the benefits, limitations, and differences between expert systems and conventional information processing tasks at the initiation of the project. A constant flow of information and updates is required to keep management involved, interested, and committed to the project.

Yet, there must be freedom for the project to change directions and use a solution other than an expert system. Flexibility is important for expert system projects because they frequently need major modifications as ideas transform and new directions are discovered, either through prototyping or conventional analyses.

Resource estimation is a critical factor influencing the success of expert system projects. It is important that the developer accurately estimates the time, staff, and financial resources required to complete the project. Another resource estimation factor is allowing for experimentation or exploration through prototyping, often both are necessary in the development of an expert system.

Measures of success for the expert system project must be clearly stated and agreed upon. Measures of success can be quantitative -- increased productivity, time savings, profits -- or they can be qualitative -- such as improved morale based on more time to do other work as a result of the time savings achieve from implementing an expert system to cover complex, but routine tasks. These measures need to be identified and defined prior to the start of the project so that they are used to guide and evaluate the expert system project.

Success factors associated with the target users of the system involve an understanding of the users' needs and a specification of the content and level of complexity of expert system's outputs. Users of expert systems do not want it to build a clock for them, they just want to know what time it is.

Who Is the User?

Identifying the intended users of the expert system leads to a clear idea of the focus of the output and its level of sophistication. Users at an entry level position will require a different focus - one that pertains directly to their task - as well as a degree of sophistication. The output should be conveyed in terms they can understand. Advanced users, on the other hand, are often better served by succinct answers that they can use as guidance. For instance, in the case of a medical application of expert systems, the difference between an entry level and advanced user might be defined by the distinctions between a medical technician and a physician.

Another success factor in determining the target output of the system is the degree of training that is necessary. A system for entry level users which requires more than a half day of training may not be a cost effective solution to an information processing problem. However, advanced users are often very busy, and a significant claim on their time for training should be "sold" to them on the basis of the results gained from using the system.

If the expert system is targeted entirely toward training, then it should focus on providing as much information to the user as possible. Training systems often try to diagnose what problems a user is having with a concept, and then set up exercises to correct the problem. Expert systems that are intended to have only incidental training benefits focus on solving the problem with a minimum amount of overhead and their output tends to be more succinct.

Explaining the System Outputs.

Justifications for the expert system's recommendations need to be clear and specific. Explanation facilities - such as the why and how queries often found in expert systems - often consist of replaying the logic used by the system to arrive at a conclusion. While this is sufficient for some applications, others require more in-depth justification including causal relationships and assumptions made. Design of the explanation facility must be tied to design of the user interface of the system.

A second facet of this issue is understanding how the output of the system is to be used. Once the users are identified and the target level of the output is set, how are the users to treat the expert system's recommendations? Are the recommendations to play the role of a checklist, an assistant, or an expert? Are users implementing the recommendation of the system immediately or do they have time to think about them before taking action? It is critical that the users of an expert system understand the degree to which they can rely upon the output of the expert system and the level of their own responsibility in using this information.

Objectives of the Definition & Design Phase.

The objectives of the Definition and Design Phase must be met in order to assure a smooth Development Stage. These objectives are listed below.

o Problem definition - the definition of the problem domain has to be refined and reaffirmed.

o Development environment selection - the success of the Development Stage hinges on careful selection of the development environment environment. The goal is to select an environment that is a good match for the problem domain.

o Delivery environment selection - the consideration of the delivery environment is an important objective of this phase. The requirements of the end users should be incorporated in the early stages of system design including evaluation of expert system shells applicable to the problem domain. Selection of development and delivery environments are linked. For instance, it makes no sense to develop a system on one host if it cannot be delivered on the host accessed by the user community.

o System design - ultimately, an overall design of the system should be achieved.

Success Factors for the Definition & Design Phase.

The development and delivery environments must be considered as integral components of the system design. Some problems fit neatly into commercial shells, while others do not. Development environments offer a variety of mechanisms for knowledge representation, control structure, and conflict resolution. These must be evaluated with care to avoid problems later during the Development Stage. Design success factors are listed below.

    The desired development environment features are validated during this phase. Various software, hardware, and technical issues should be addressed in the selection of a specific product.

    The first consideration in selecting a development environment is determining the hardware and software currently available to the user base. This impacts, and may constrain, the choice of both development and delivery environments.

    The developer can explore the wide variety of PC-based shells first to try to find a match with the problem domain. Shells can provide a quick solution to many problems. Do not attempt to use complex languages without proper training, appreciation for the difficulty of the task, sufficient time for development, or the availability of experienced programmers.

    If a shell has been selected as a development environment, be sure that the tool is affordable, easy to install, and easy to use. An overly expensive tool dilutes the benefits derived from the system. Choose a tool that can accommodate the problem. The tool must be able to handle the necessary data, and which can also easily import or export data. A superior shell can be modified to provide additional features.

    Special development hardware may be required to run the tool. The additional cost must be considered in the selection process.

Integration with desired data sources, programs, and output interfaces should be readily accommodated. Establish reasonable goals for satisfactory performance in terms of speed and memory use. Migration to a suitable delivery vehicle should be easy and inexpensive.

Verification and validation methodologies must be determined at this stage. Verification techniques are the methods used to determine that the expert system has been built correctly.

Verification techniques for the software, knowledge base, and interfaces should be derived from the expert system's functional requirements. Steps taken and methods used to verify the expert system need to be mapped from the components up through interfaces and software module interactions. The operational points of the expert system that need to be validated - such as scope and effectiveness - are identified at this point.

Validation techniques are the methods used to determine that the expert system conforms to the functional requirements and can be used as intended. The validation techniques should be identified in the Definition and Design Phase. Issues such as the need for external experts, and types and location of test cases should be thought out.

When establishing a design for the delivery environment, there are several things to be aware of. These include: (1) the environment's flexibility, availability, and compatibility with equipment currently being used, (2) licensing fees required for distribution of the system. Be sure ask the vendor about licensing fees when selecting a shell in terms of producing "run time" versions. (3) graphics requirements for the production system, (4) evaluation of any potential problems caused by excessive data migration from the mainframe, or any larger host, to a PC. If it appears that there will be excessive data migration, perhaps it is a case when a mainframe implementation should be used, (5) the migration from the development environment to the targeted delivery environment must be technically possible. Careful selection of development and delivery environment will avoid problems.

The end user interface is an important portion of the expert system because it is the primary means by which the systems communicates with the user. User involvement in selecting the desired features should be included as early as possible to ensure the success of the system. Some issues to address include: (1) incorporating interesting, user-friendly screens and system features, color always helps (2) providing convenient data input for the end user without the advanced editing functions needed only in development, (3) ensuring adequate system response time, especially for calculation-intensive applications, and, (4) creating query interface capabilities that support flexibility and sophistication. You can count on end-users attempting things with the system for which it was not riginally intended. How it responds to these challenges may influence the acceptance of the system.

Objectives of the Development Stage.

Key objectives in the Development Stage are to document key issues to be borne out during prototyping. The problem description, functional description, and knowledge base are refined during development of the prototypes. Now is the time to increase product visibility. Also, maintaining management commitment is a key objective of the Development Stage. Another objective is to codify the knowledge base and maintain its accuracy through testing.

Success Factors of the Development Stage.

There are several areas that are crucial to the success of the Development Stage. These factors are grouped below according to the area they address.
    Knowledge engineering begins in the Concept Phase. All the work which has been down since then pays off int he development stage. The knowledge engineers should devise a method to resolve conflicts among multiple sources of expertise with differing specialties. Methods for dealing with conflicts must be developed and implemented now. It is wise to coordinate programmers and knowledge engineers, but keep in mind that there may be organizational separation and differences in background.
    The parallel of a configuration management plan used in conventional system is to the use of a "knowledge management plan" used for expert systems. It is used to keeps track of all the rules or other "expertise" used in each successive iteration of the system.
    Knowledge engineers should have a strong computer background to facilitate communication with programmers. Knowledge engineering is most effective when proven knowledge acquisition methods are used. These include :

o Unstructured interviews,
o Structured interviews,
o Observation,
o Interruption analysis,
o Constrained-processing tasks,
o Questionnaires, and
o Decision trees and decision tables.

    Many good techniques are available and are described in the literature. The knowledge engineer should apply one or several structured and unstructured techniques to document the expert's domain knowledge. The knowledge engineers and expert(s) should think out all of the ramifications of the rules. It is a good idea to periodically recheck the accumulated knowledge in order to better refine it.
    Knowledge engineers need not be experts in the field; too much domain knowledge has a risk of producing biases in the process toward the knowledge engineers and away from users. Select knowledge engineers who are familiar with the domain, but are not necessarily expert in it. Also, there is a question of how much knowledge engineering should be imposed on a chemical engineer. The answer may be "not too much," or a company may lose a valued employee to the expert system consulting firm.
    Knowledge engineering time estimates should be well thought out with respect to reaching the appropriate depth of knowledge. There should be a constant reminder to the project manager to plan for contingencies and expect repeat sessions with experts and users.
    Following are several methods that contribute to a successful prototype.

o Use rapid prototyping with frequent interim deliverables and decision points.

o Discard the prototype if a better design approach is discovered. The purpose of a prototype is to firm up the design specifications. It helps to insure that management understands the idea of "disposable code."

o The design should be modular to show all lines of reasoning and specific functions the expert system is required to perform. The prototype should also be attractive as a demonstration vehicle to gain the support of users and management. The development team should accurately define and adhere to the system development stages. This will allow the team to anticipate the long-term impact of schedule changes.

It is important to identify and validate external interfaces as early as possible. This includes inputs and outputs, and other programs, and algorithms - separately from the knowledge bases. Validation should be done frequently and continually. There is a tendency to ignore validation until the end of the Development Stage. Ideally, it should be done after the addition of each rule, but later can be reduced to the end of each session. Expert system validity relies heavily on the validity of accessed data. Issues to be considered when validating the system include the knowledge base, recommendations, justification, rationale, type of inferencing and what happens when the system is given incomplete data.

It is not always possible to test all possible rule outcomes. This is often the case for larger systems, e.g, +500 rules. This emphasizes the importance of keeping a "trace" of each session and the need for software maintenance in the operational phase.

Testing is the most important part of the Development Stage because there may be potential conflicts in the knowledge base. A thorough analysis should be performed on a routine basis throughout development and upon completion of development. Make sure to adequately test critical components and any examples or rules generated by inductive systems. Thoroughly retest entire knowledge base when changes are made. The process of testing expert system shells is easier than testing systems written in LISP or PROLOG because the inference engines have already been thoroughly tested in a shell.

Objectives of the Implementation Stage.

The objectives of the Implementation Stage are to complete testing of the expert system and prepare it for distribution. This includes a beta test of the system by potential users and a revision of the system following the test. In preparation for distribution of the system, a training program and documentation are completed, and users are registered. The expert system itself is also prepared for distribution through final debugging and copies of the run-time version are produced.

Success Factors of the Implementation Stage.

Most of the success factors involve the transfer of the system to users in the field and the initial use of the system. In order to produce a successful expert system, it is important to:

o Provide useful, readable documentation

o Provide organized training

o Provide technical support

o Ensure that hardware in the field is adequate to support the expert
  system software

o Complete licensing and run-time versions

o Maintain management commitment during the distribution of the system.

Most of these factors can be covered through planning and coordination prior to this stage.

Management commitment during this stage can be ensured if, early
in the life cycle, management is made aware of the fact that they will
play a large role in the Implementation Stage. They should know that
they will be responsible for the scheduling and planning involved in
this stage, as well as the actual distribution of the system and all
of the issues accompanying the distribution process. For instance,
management commitment is needed to make the time of users of the system
available to attend system training classes.

Good documentation can be produced through a quality assurance
system, requiring reviews by the developers of the system, as well as
management. A draft of the documentation must be provided to the beta
users for their comments on its clarity and comprehensiveness.

The planning of the training process should be supervised by management, who will work with the developers and the trainers. A Technical
support team should be organized before the system is distributed.
Their duties should include manning a support hotline and providing
help to the users. The technical support team should include individuals
who were involved in developing the expert system, writing the
documentation, and providing training sessions. In order to avoid tying
up expensive talent with the help line, a 4-hour turnaround in answering
queries may be acceptable. Alternatives include the use of electronic
mail or PC-based bulletin boards.

Licensing and run-time issues should have been confronted early
in the life cycle. During the Implementation Stage, it should be
necessary only to confirm with the vendor of the shell what steps should
be taken to distribute run-time copies of the expert system. The matter
of who will pay for the run-time copies should have been resolved by
now. Problems with hardware in the field can be avoided if the users
are given enough time to acquire equipment required to run the expert
system.

Objectives of the Operations Phase.

The first objective of the Production Stage is to use the capabilities
of the system to solve the information management problem by delivering
the system to the users. Proper use of the system will require user
training in the capabilities and limitations of expert systems.

The second objective is to identify potential changes needed to
ensure that the system and data continue to solve the information
management problem. Changes may take the form of routine maintenance

or may constitute enhancements to the system or databases. Careful management of the knowledge base, including revalidation after each change, will be required to ensure continued satisfactory performance.

The third objective is to develop and implement maintenance changes and minor enhancements. All maintenance and minor changes are controlled through configuration and knowledge management baselines. If a system is to be taken out of service, users must be notified as well as the support team.

Success Factors of the Operations Phase.

In this users should not become overly reliant on an expert system's recommendations. Because the advice offered by the system is so similar to the human expert's recommendations, the advice may be accepted as infallible. This is particularly dangerous when the system is working with incomplete or incorrect information. Users should maintain an attitude of skepticism in proportion to the consequences which result if the system renders incorrect advice.

The organization should be cognizant of changes that affect the system's performance and should not treat them too casually. Relevant changes such as development of new techniques or expertise, and enhancements in hardware and software may affect the system in discrete pieces or synergistically. Formal evaluations of all perceived changes and their impact on the system are required.

Translating changes into modifications to the knowledge base will require an iteration of knowledge engineering with the expert. Depending on the extent of the changes, this iteration could almost be a mini-project development effort.

Over time, the problem that the system was developed to address may cease to exist or may be subsumed in a larger problem addressed by another system. "Pride of ownership" can inhibit shutting down a system that is no longer required. This is obviously wasteful and inefficient. All programs and systems should be viewed objectively.

If a system is shut down, it is important to retain components of the system that may be useful at a later date. In general, knowledge is always valuable. Knowledge in automated form is easily archived and retrieved as needed. However, it is also easy to overcompensate for the first by retaining everything. If an information management problem ceases to exist, significant portions of the solution to the problem probably can be discarded. Large systems may consume considerable physical and logical storage space; both are expensive and should be used efficiently.

Conclusion

This paper has reviewed the objectives and success factors for developing expert systems, portraying the rapid prototyping cycles of expert systems against the framework of traditional life cycle management practices.

Acknowledgments

The author gratefully acknowledges the assistance and support of the following individuals in the development of this paper and the program described herin. Any errors are the responsibility of the author.

- John Convery, Acting Deputy Director, Risk Reduction Laboratory, US EPA, Cincinnati, Ohio.

- Dan Greathouse, Operations Research Specialist, Risk Reduction Laboratory, US EPA, Cincinnati, Ohio.

- Asa R. Frost, Jr., Director, Information Management Staff, Office of Solid Waste & Emergency Response, US EPA, Washington, DC.

- Henry Horsey, Research Director, Center for Advanced Decision Support Water Environmental Systems (CADWES), University of Colorado, Boulder, Colorado (formerly with CDM Federal Programs, Chantilly, Virginia).

- Jerry Feinstein, Vice President, ICF-Phase Linear Systems, Fairfax, Virginia.

- Randy Manner, Principal, Expert Systems Group, American Management Systems, Arlington, Virginia.

Literature Cited

1. Harmon, P.; King, D. Expert Systems: AI in Business; John Wiley and Sons, New York, NY, 1985.

2. Hayes-Roth, F.; Waterman, D.; Lenat, D. Building Expert Systems; Addison-Wesley, Reading, MA, 1983.

3. Waterman, D. A Guide to Expert Systems; Addison-Wesley, Reading, MA, 1986.

4. US EPA, Office of Solid Waste & Emergency Response, "System Life Cycle Management Practice Paper on Expert Systems;" Washington, DC, 1989.

RECEIVED January 22, 1990

Chapter 3

Verification and Validation of Environmental Expert Systems

Mark Stunder

GEOMET Technologies, Inc., 20251 Century Boulevard, Germantown, MD 20874

This chapter provides an overview of expert-system verification and validation (V&V) techniques. Several methods are presented. First, many of the conventional software V&V techniques such as requirements analysis and unit testing can be applied to expert-system development. Second, an expert-system developer can use automated tools to test rule consistency and structure. A more viable alternative, however, is for the developer to create his own set of consistency and completeness tests. Finally, a developer should rely on qualitative judgment to determine the validity of a knowledge base. This judgment could include expert opinion as well as specialized tests designed to determine knowledge-base certification. The chapter suggests that methods should be combined into an optimal mix in order to best undertake V&V.

The goal of every software procurement effort, whether for the government or private industry, is to obtain a validated, working, usable, and cost-effective system. When expert systems (ES) are included in such a software effort, the overall complexity of systems increases with a corresponding increase in the importance, difficulty, and effort required for verification and validation of the delivered system. Expert systems present problems because they often fail to explicitly represent the reasoning, knowledge, and thought that went into their design. Often the code itself does not reveal its relationship to higher-level linkages. This hampers a programmer's or knowledge engineer's diagnosis of the causes of an error and the locating and modifying of all relevant code.

Verification is the review of system requirements to ensure that the system has been built to specifications. Thus, verification involves communicating activities and results of project activities. This is done through documentation and verifying that certain steps have been taken.

0097–6156/90/0431–0039$06.00/0
© 1990 American Chemical Society

Validation involves determining that the system performs with a reasonable level of accuracy. Validation is accomplished through test and evaluation of ES software and integrated hardware. Validation thus ensures that the capabilities that have been specified in the ES requirements have been exercised and meet levels acceptable to the user. Thus, without a true V&V methodology, much time is lost in the evolutionary expert system development process.

It may seem strange to the reader that a chapter on V&V appears early in a book on environmental expert systems; however, this is done purposely. To achieve success in the fielding of environmental expert systems, we must understand that V&V is part of the overall expert system life cycle. In fact, how V&V will be accomplished should be part of the expert-system design specification, statement of work (in the case of government contracting), or spelled out in a task assignment. This means that V&V needs to be openly discussed by the expert system development team and the client on day one. Too often have we seen expert systems being produced without formal verification (knowledge-base certification) or without validation. This is particularly common in the environmental area where very few expert systems have gone through rigorous evaluation.

The purpose of this chapter is to describe three methods that are available for environmental expert system V&V. These methods are:

(1) Adaptation of conventional V&V methods to expert system V&V
(2) Fully automated procedures and ES developer testing tools
(3) Specific knowledge-base verification procedures (qualitative review).

It is hoped that environmental expert-system developers can utilize these techniques to provide reassurance to clients on the quality of their expert-system efforts.

Method I: Conventional V&V Applied to Expert Systems

The first method draws on the guidance available from the conventional software V&V literature and attempts to treat expert system V&V like conventional V&V whenever possible. This approach has been suggested by several authors (see for example Jacob and Froscher 1986). In the conventional approach, software V&V is like the quality assurance of any other product: it refers to the procedures used to ensure that the product meets its specialized development requirement. For example, the Department of Defense (DOD) requirements for software development found in DOD-STD-2167A call for software development activities, products, reviews/audits and baseline/developmental configurations.

That document specifies a default development cycle which has become known throughout the government as the "waterfall chart." This chart is shown in Figure 1. The Roman numbers on the chart indicate the five phases of the conventional life cycle which are:

(I) Requirements Development and Analysis
(II) Design

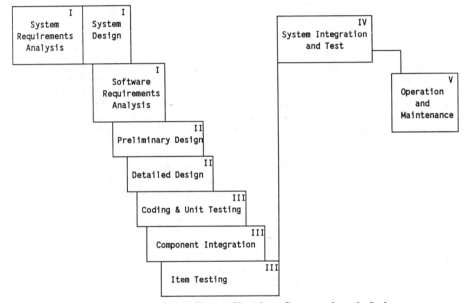

Figure 1. Waterfall Chart Showing Conventional Software Development Cycle

    (III)   Encoding and Testing
    (IV)   Integration and Installation
    (V)   Operations and Maintenance.

The requirements phase involves definitions and limitations of the problem and creation of plans (project plans, V&V plan etc.). The design phase involves development of design specifications and logical processing requirements. The encoding and testing phase involves actual software code generation and software unit testing as well as any external interface testing. The integration and installation phase involves placing the system in operation and ensuring that all components and the system as a whole have been tested properly. Finally, the operations and maintenance phase involves checking the system for any problems after installation and modifying the software accordingly.

Many non-DoD government and private groups have also adopted the general "waterfall approach" as their overall conventional software development cycle or methodology. For example, Wilburn (1983) for Westinghouse, Powell (1982) for the National Bureau of Standards and Bryant and Wilburn (1987) in NUREG/CR-4640, show how verification is imbedded throughout the waterfall cycle of software development. Requirements specification analysis, functional specification/ detailed software design, coding/software generation and of course the integration and testing stages all contain V&V elements.

Table I shows how the "waterfall chart" phases can utilize various conventional software V&V methods. Unfortunately many people (Weaver 1989; Miksell 1989) believe that verification specifically belongs in the true testing phase only. In reality, the most efficient use of verification is throughout the software cycle and particularly during the requirements analysis.

Verification within testing itself, however, can be divided into three parts called:

- Unit testing
- Integration testing and
- Formal qualification testing.

Unit testing is the testing of the smaller identifiable software components (units). Integration testing is performed to demonstrate that units and higher-level components of the system work together. Formal qualification testing is another term for validation and is performed to formally demonstrate that the software meets its established requirements.

Most expert-system development efforts will also follow the five phases of the conventional life cycle with heavy emphasis on knowledge acquisition and representation in the design phase. Several of the V&V procedures outlined in Table I for conventional software life cycle can be applied in ES V&V. Table II provides a list of some applicable procedures. Many procedures are simply a part of _any_ software development effort, but are many times overlooked in ES efforts. For example, ES developers oftentimes do not develop adequate requirements documentation. They have difficulty in tracing knowledge representation structures (rule paths, frames, etc.) back to design documents which in turn makes

Table I. Conventional Software V&V Methods Related to
Conventional Software Development Cycle

Waterfall Cycle Phase	Possible V&V Method
(I) Requirements Analysis	• Tracking of requirements • Develop testable require-ments • Use of structured methods (break big pieces to smaller pieces) • Conceptual model develop-ment/verification
(II) Software Design	• Use of assumptions as check points; violations cause known items to occur • Structured programming techniques (aids in V&V) • Complete design documen-tation • Tracing of design back to requirements
(III) Encoding and Testing	• Peer review (overall check) • Team development (team check against itself) • Use of approved standards (format like modules alike) • Various code and data flow analyses • Design parallel modules (error in one could be an error in another) • Unit testing • Integration testing • Formal qualification testing
(IV) Integration and Installation	• Systems testing/simulated operating environments/ documentation • Acceptance testing (after installation) • Summary of verification performed • Appropriate validation and acceptance
(V) Operations and Maintenance	• Checks of performance requirements • Verify any modifications made

Table II. Use of Conventional V&V Procedures in ES Development

Development ES Phase	Example V&V Procedures
(I) Requirements Analysis	• Conceptual modeling/flowcharting • Tracking requirements (particularly if ES will be large)
(II) Design	• Knowledge-base design to requirements check • Modular programming strategy (if ES design permits)
(III) Testing	• Unit test chunks of ES code • Test integration of outside (i.e., LOTUS) programs
(IV) Integration and Installation	• ES reproducibility check (mirror expert?) • Acceptance testing
(V) Operation and Maintenance	• Verify any modifications mode

knowledge base verification difficult.   Emphasis on traceability
could, therefore, alleviate a potential verification stumbling
block.

Stunder (1986) has shown that two issues arise from the direct
application of conventional V&V techniques to expert system V&V.
These issues are:

- Establishment of clear acceptance criteria
- Architectural items.

Acceptance criteria for expert systems depends largely on the
function of the system to be developed.   This relates to the need
for clearly defining requirements.   For example, the Thermal
Performance Advisor System developed for the Electric Power Research
Institute and used in nuclear powerplants relies on criteria
centered around efficient plant performance.   The Zeus
meteorological system, devel-oped by GEOMET, Inc., for the United
States Air Force (Sletten et al. 1988), built acceptance criteria
around site flight profiles and acceptable weather forecast
parameters as a means of verification and validation.   GEOMET's
Underground Storage Tank regulatory system on the other hand used
clearly established EPA regulations and procedures in establishing
acceptance criteria.   In all cases, acceptance criteria were
documented in the requirements phase.

Many times, the acceptance criteria also depend on the type of
knowledge to be captured.   Sometimes an expert-system approach is
chosen as a convenience to incorporate "hard" knowledge, such as
inflexible rules involving regulatory requirements (Stunder and
Hlinka 1989).   In such cases, the expert system approach has no
direct affect on the acceptance criteria; they are no different than
for conventional software.   Frequently, however, an expert system is
chosen to encode knowledge that has a "soft" component.   This means
that knowledge needs to be gathered for the system from experts or
from reference material.

A second area affecting <u>direct</u> applicability of conventional V&V
techniques to ES V&V is the architectural structure of expert
systems.   First, guidelines for assigning architectural levels in
conventional software do not necessarily apply in expert system
design.   For example, in conventional software development, units
are defined for testing purposes; in ES development, units may be
harder to define because of the dependency of rules, objects, etc.
Secondly, in rule-based systems, rule interactions are frequently
difficult to predict.   Oftentimes, the ES code does not reveal its
relationship to higher level linkages.   A programmer may have a
difficult time finding the cause of an error and modifying all
codes.

Consistency and completeness checking algorithms are available
(Nguyen et al. 1985) which can point out simple examples of
conflicting rules and missing rules.   Some of these rule problems
are shown in Tables III and IV, but there can be additional subtle
interactions across a rule base.   There are five possible ways that
rules can be inconsistent.   These include:

- Redundancy
- Conflict

Table III. Common Consistency Problems

Redundancy - two rules have the same antecedent (IF part) and their conclusions (THEN part) contain identical actions/clauses, e.g.

        Rule 1: IF (SUB_SYSTEM1(STATUS) = ABNORMAL)
                 THEN NOTIFY(OPERATOR)

        Rule 2: IF (SUB_SYSTEM1(STATUS) = ABNORMAL)
                 THEN NOTIFY(OPERATOR) AND SHUT_DOWN(SUB_SYSTEM1)

Only Rule 2 is necessary.

Conflict - two rules have the same antecedent, but their conclusions are contradictory, e.g.

        Rule 1: IF (SUB_SYSTEM1(TEMP) > 140)
                 THEN SUB_SYSTEM1(STATUS):= ABNORMAL

        Rule 2: IF (SUB_SYSTEM1(TEMP) > 140)
                 THEN SUB_SYSTEM1(STATUS):= NORMAL

Subsumption - two rules have the same conclusion, but the antecedent(s) of one is contained within the other, e.g.

        Rule 1: IF (SUB_SYSTEM1(TEMP) > 140)
                 THEN SUB_SYSTEM1(STATUS):= ABNORMAL

        Rule 2: IF (SUB_SYSTEM1(TEMP) > 140) AND
                    (SUB_SYSTEM1(VOLTAGE) = 0)
                 THEN (SUB_SYSTEM1(STATUS):= ABNORMAL

Rule 2 is unnecessary.

Unnecessary IF rules - two rules have contradictory clauses in otherwise identical antecedents, and identical conclusions, e.g.

        Rule 1: IF (SUB_SYSTEM1(STATUS) = ABNORMAL) AND
                    (SUB_SYSTEM2(STATUS) = ABNORMAL)
                 THEN NOTIFY(OPERATOR)

        Rule 2: IF (SUB_SYSTEM1(STATUS) = NORMAL) AND
                    (SUB_SYSTEM2(STATUS) = ABNORMAL)
                 THEN NOTIFY(OPERATOR)

The conclusion, NOTIFY(OPERATOR), is based strictly on the "identical" portion of the antecedent, (SUB_SYSTEM2(STATUS) = ABNORMAL). These two rules should be combined into one rule, eliminating the (SUB_SYSTEM1(STATUS) = ...) clause.

Circularity - a set of rules forms a cycle, resulting in the Expert System equivalent of an infinite loop.

Table IV.   Common Completeness Problems

Unreferenced attribute values - legal values which may be
assigned but result in no further processing, i.e., no other
rules refer to that value, e.g.

      IF...THEN SUB_SYSTEM1(STATUS):= QUESTIONABLE

but no other rules refer to QUESTIONABLE status.

Illegal attribute values - a rule refers to an attribute's value
where the value is not a legal value, e.g.

      IF (SUB_SYSTEM1(STATUS) = DESCENDING) THEN

where DESCENDING is not a legal value for subsystem status, only
for the satellite.

- Subsumption
- Unnecessary ifs
- Circularity.

Examples of each problem are given in the table; however, circularity and conflict appear to be the most prevalent of the problems. Circularity involves a set of rules firing and resulting in a repetitious answer. The key to stopping circularity is to find the proper point in the rule path to branch off (without destroying another portion of the code). The process can be tedious. Conflict usually results from poor design or knowledge representation where rules provide contradictory conclusions.

Completeness problems typically involve rules that are ends in and of themselves. For example, a rule may reference an attribute value which does not result in future processing. Similarly, a rule may reference an illegal value, whereby the rule fails.

Rule order may influence both the operation (and therefore the results) and the execution time of a rule base. Therefore it is necessary to test the expert system under a demanding variety of scenarios.

Many of the conventional V&V techniques within the conventional software life cycle can either be directly applied to BS development or can be modified to handle the somewhat instructured nature of an ES. Formal unit testing used in conventional software development can also be applied by looking for items such as consistent and complete rules.

Method II: Fully Automated and ES Developer Testing Tools

Fully Automated Methods. The second approach to expert system V&V has been the use of verification tools. These tools generally analyze the source code for the expert system. This analysis does not involve executing or exercising the system. Verification tools for expert systems are similar to those of conventional software. These tools process the source code and in some cases descriptions of the source code looking for problems that can be determined mechanistically by a program. Conventional tools can detect:

• Uninitialized variables	• Unreachable sections
• Type mismatches	of code
• Number of lines per module	• QA standard
• Redundant code	violations
	• Proper commenting

Tools for rule-based expert systems (as well as manual methods) should evaluate the consistency and completeness of the rules. The TEIRESIAS program (Davis 1976) linked to the MYCIN infectious disease system was one of the first attempts to develop an automated verification tool. Later work by Suwa et al. (1982) for the ONCOCIN (clinical oncology) system examined a rule set as it was read into the system. This rule checker assumes that for each combination of attribute values appearing in the antecedent a corresponding rule exists.

The LES system described by Nguyen et al. (1987) is a generic rule-based expert system building tool which has an extensive

checker called CHECK.   Redundant and conflicting rules along with
many other types of problems can be uncovered with CHECK.
The problem with automated tools is that many times they are
software/system dependent.   For example, CHECK has been ported to
the Automated Reasoning Tool (ART) framework but to little else
(short of a MYCIN type structure).   Thus, automated tools are not
necessarily available for off-the-shelf applications in shell,
language or tool expert system development.   Cost-to-benefit-type
trade-off studies need to be undertaken before fully automated tools
are fully recognized as a (partial) means of expert system V&V.

Expert System Developer Testing Procedures.   An expert system developer can generate his/her own code for testing various elements of
the ES code without going to fully automated self-contained
procedures.   Borrowing from the conventional software side, three
methods can be used for code testing.   They are:

- stubs
- drivers
- simulators.

A stub involves a set of code which produces known results.   For
example, a stub imbedded in ES code could contain "canned" facts
rather than executable rules.   A stub could be written to determine
whether the output from a module is correct.   The stub may be as
simple as a "canned" temperature value as part of a rule.   If the
rule path is executing properly, that "canned" temperature value
would be returned to the module.   Stubs are usually utilized early
in the ES software development cycle where linking rules, objects or
frames are not available, but pieces or units of ES code are
available.   Stubs are also occasionally inserted in the integration
phase to force a known result or to verify a portion of the
knowledge-base.
Drivers provide another means of testing ES code.   They are much
larger in scope and essentially can drive the software unit under
test.   A driver can exercise a portion of the overall program over a
full range of possible outputs, thus allowing for determination of
error-bound crossings (e.g., unrealistic values) or lack of code
execution.
A third technique that a developer can use to test ES code
involves simulation.   Simulators can test entire systems or parts of
systems and are similar to drivers.   A simulator approach is more
functionally oriented than a driver in that a simulator usually will
be utilized to test a call for a conventional language or package
(e.g., LOTUS), whereas a driver is interested in values or a
specific ES element.   Simulators can also generate data input for
real-time testing of ES code.   Use of a simulator test approach,
particularly just prior to integration and installation, allows for
initial data and information to be generated for Method III which is
the qualitative knowledge-base review approach.

Method III: Qualitative Knowledge-Base Review

The qualitative review of knowledge bases is a major issue facing
expert-system developers.   It is sometimes referred to as knowledge-

base certification.   This is because of expert disagreement on
knowledge-base content.   Many authors suggest that knowledge-base
expert disagreement could be downplayed by the proper selection of
expert-system applications.   Many expert systems are unfortunately
developed in areas which are still in the basic R&D stage and are
not ready yet for widespread use.

In determining expert opinion on ES output, a variation of the
classic Turing Test (Stunder et al. 1988) can work with a group of
experts.   Test design involves presenting a series of simulated
cases to a group of experts and to the expert system.   This is done
to verify not only that the entire expert system responds in a way
consistent with human performance, but that each piece of the
knowledge base executes correctly.   [Ideally it should be impossible
for an observer reviewing test results to determine which set of
results comes from interaction with the expert system (Jacobs and
Chee, 1988).]   Independent review of the knowledge base is possible
by using various acceptance levels, human performance checks and
selected groups (Stunder et al. 1988).   The use of a simulator
technique described under Method II can generate scripts which when
reviewed against expert opinion can provide useful knowledge-base
review.

In qualitatively reviewing the knowledge-base, it is sometimes
difficult to determine the minimum competency level of the system in
order to even undertake a Turing Test approach.   Thus, defining
minimum environmental expert system competency is a difficult task
which should be done in the software requirements phase.   This is
where many expert systems fail.   Criteria are usually not developed
to assess expert system performance in the early stages.
Quantifiable criteria such as "average accuracy" will aid in any
knowledge-base review process, but judging which rules meet
competency criteria is difficult.

The recent experience of Lal and Peart (1989) with the FARMSYS
agricultural expert system, for example, indicates that the methodology of a small group of experts meeting together to evaluate and
critique a system offers a good practical alternative for verifying
and validating a system.   Similarly, Stunder et al. (1986) used a
group-oriented methodology to not only validate actual environmental
system performance but to verify the physical concepts imbedded
within the knowledge base.   The degree of qualitative review will
vary for each ES development.   Developers should carefully consider,
however, how experts can review the expert system in an unbiased
method.   The Turing Method provides a means of undertaking such an
evaluation.

Summary

The proliferation of environmental expert systems means that more
attention must be spent on how these systems are fielded.   Proper
V&V of expert systems ensures that a client's contract dollars are
well spent on the overall system effort.   It is especially important
to include V&V in all design specifications.

This chapter has outlined three general approaches for
undertaking V&V.   There is no one single approach.   Instead, all
approaches and techniques should be used in some form during an
entire expert system life cycle.   As more environmental expert
systems are fielded, it is hoped that the writing of V&V
requirements becomes mandatory throughout the expert system life
cycle process.

LITERATURE CITED

1. Bryant, J.L.; Wilburn, N.B. <u>Handbook of software and quality assurance techniques applicable to the nuclear industry</u>; NUREG/CR-4640, 1987.
2. Davis, R. Applications of meta-level knowledge to the construction, maintenance, and use of large knowledge bases, Stanford University, Dept. of Computer Science, Ph.D. dissertation, 1976.
3. Jacob, R.J.; Froscher, J.N. <u>Developing a software engineering methodology for knowledge-based systems</u>, TR9019, Naval Research Laboratory, 1986.
4. Jacobs, J.M.; Chee, C.W. <u>Specification of expert systems-testing</u>, TR 88-303, HQ Space Division USAF, 1988.
5. Lal, H.; Peart, R.M. <u>Engineering farm knowledge for a seamless decision support system</u>, 1989.
6. Miksell, S.G. <u>Operational evaluation of an expert system: the FIESTA approach.</u> Ann. Conf. Int. Assoc. of Knowledge Eng., University of Maryland, 1989.
7. Nguyen, T.A.; Perkins, W.A.; Laffey, T.J.; Pecora, D. Checking an expert system's knowledge base for consistency and completeness. <u>In Proceedings: Ninth Joint Int. Conf. on Artificial Intelligence</u>, Menlo Park, California, 1985.
8. Nguyen, T.A.; Perkins, W.A.; Laffey, T.J.; Pecora, D. Knowledge-Base Verification. <u>AI Magazine</u> 1987, <u>8</u>(2), 69-75, 1987.
9. Powell, P.B. <u>Software validation, verification and testing: technique and tool reference guide</u>, NBS SPI Pub NBS-SP-500-93, National Bureau of Standards, Washington, DC, 1982.
10. Sletten, T.N.; Stunder, M.J.; Lee, S.M. <u>Use of environmental expert systems.</u> Army Information Systems and AI Conference, American Defense Preparedness Association, El Paso, Texas, 1988.
11. Stunder, M.J. <u>Managing AI-environmental applications.</u> First Artificial Intelligence Research in Environmental Sciences Conference, Environmental Research Laboratory, NOAA, Boulder, Colorado, 1986.
12. Stunder, M.J; Hlinka, D.J. <u>Applications of hypertext systems in the environment</u>, Third Artificial Intelligence Research in Environmental Sciences Conference, Washington, DC, 1989.
13. Stunder, M.J.; Pelton, D.J.; Lee, S.M. <u>Zeus II: a multipurpose environmental expert system</u>, 55th Operations Research Society Meeting, Montgomery, Alabama, 1988.
14. Stunder, M.J.; Pelton, D.J.; Sletten, T.N. <u>A meteorological knowledge-based expert system</u>, First Artificial Intelligence Research in Environmental Sciences Conference, Environmental Research Laboratory, NOAA, Boulder, Colorado, 1986.
15. Suwa, M.; Scott, A.C.; Shortliffe, E.M. An approach to verifying completeness and consistency in a rule-based expert system, <u>AI Magazine</u> 1982, <u>3</u>(4), 16-21.
16. Weaver, S.J. <u>Conventional testing and knowledge-based systems</u>, Ann. Conf. Int. Assoc. of Knowledge Eng., University of Maryland, 1989.
17. Wilburn, N.P. <u>Guidelines-software verification.</u> HEDL-TC-2425, Westinghouse Hanford Company, Richland, Washington, 1983.

RECEIVED February 12, 1990

Chapter 4

Neural Networks and Environmental Applications

Joseph Schmuller

Expert Systems Team, CDM Federal Programs Corporation, 13135 Lee-Jackson Memorial Highway, Fairfax, VA 22033

Neural network models are gaining popularity in a variety of areas, and are currently getting a lot of publicity. This paper is a tutorial introduction to these models, and is intended for people with little or no background in the field. We begin with the fundamental concept of interconnections among simple computational elements, examine a simple neural net model, and discuss training and validation of neural networks. We outline differences between neural net models and traditional expert systems, and we present ideas for using them together. Major players in the neural net field are enumerated, as well as the roles they play. We list successful applications of this technology, and we indicate several potential environmental applications.

A neural network is a piece of hardware or software (or a combination of the two) that simulates what we think we know about how the brain works. The operative word is "think"; to say that such models do things exactly as the brain does would be much too presumptuous at this time, as there are still a great many mysteries concerning the brain. To avoid any implied presumption, some researchers in this field prefer terms like parallel distributed processors, connectionist systems, or collective decision circuits.

Neural network models -- and brains -- contain sets of elements, each of which is computationally simple. The elements, called "neurons", are highly interconnected to one another; in the human brain there are about 100 billion neurons, and each one is connected to about 10,000 other neurons. Neural network models often contain large numbers of simulated neurons, but not as many as are in the brain. For the remainder of this paper, we will refer to simulated neurons as "units".

0097–6156/90/0431–0052$06.00/0

A Closer Look

How does a networked arrangement of units solve problems? Let's look
at a simple network.

Layers. Figure 1 shows 15 units on the left connected with 5 units
on the right. Let's suppose that the units on the left take a pattern
of stimulation from the outside world, and that this pattern is then
represented in these units. Let's refer to the 15 units as the "input
layer". Each of the 5 units on the right is connected to every unit
in the input layer; thus, the activity of these units depends on the
pattern of input. We'll call these 5 the "output layer".

Stimuli and Responses. Imagine that each unit in the input layer
corresponds to a cell in a 5 X 3 matrix, and that each of these cells
can have the value "1" or "0" (corresponding to "on" or "off"); each
unit in the output layer corresponds to an uppercase version of one
of the first 5 letters of the alphabet. Any pattern on this matrix can
be represented by a vector of 15 numbers, each of which is either 1
or 0. The problem that this network has to solve is this: given a
vector of 1's and 0's presented to the input layer, which of the
letters A through E does the vector correspond to?
    Figure 2 illustrates such a 5 X 3 matrix and 5 upper-case letters
that could be constructed with this matrix. In terms of 1's and 0's,
the matrix's vector for the letter A is

$$[0,1,0,1,0,1,1,1,1,1,0,1,1,0,1]$$

We want only the first output-layer unit (which corresponds to
"A") to fire if this pattern is presented. That is, in terms of 1's
and 0's, we would want the output-layer's response vector to be

$$[1,0,0,0,0].$$

Unit 8 in the input layer corresponds to the center in A, B, and
E. If this unit is "on", the network has evidence that the input
pattern is not C or D. If this unit is "off", the network has
evidence that the input pattern is not A, B, or E. We can make
similar statements about other units and the letters they are
associated with and the ones they eliminate. If a unit is associated
with a letter, we will attach a positive weight to its interconnection
with the output-layer unit corresponding to the associated letter.
If a unit's firing (or being "on", or having the value "1" in the
input pattern) eliminates a letter, we'll attach a negative weight
to its interconnection with the letter's output-layer unit. Each
output-layer unit, then, has a set of weights, and each weight
corresponds to a connection with an input-layer unit. Each weight is
a positive or negative number, and each input is either a one or a
zero.

Activation of a Unit. What does a unit (in this case, an output-layer
unit) do with these inputs and weights? In simple networks like ours,
a unit j takes an input (x_i) and multiplies it by the weight of the
interconnection (w_{ij}) through which the input came. It does this for
all its inputs, and then adds these products together. That is,

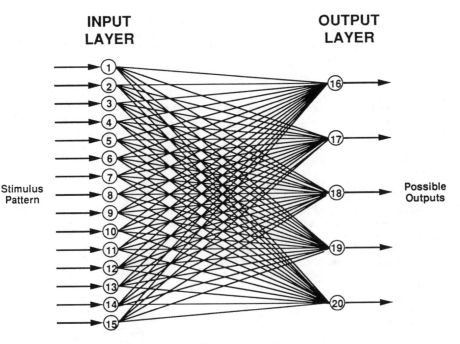

Figure 1. A Simple Neural Network

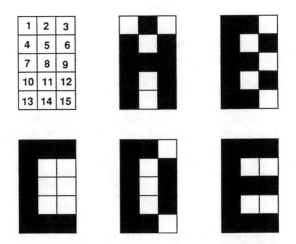

Figure 2. Stimulus Input Matrix and Uppercase Letters

$$A_j = \text{SUM } x_i w_{ij} \qquad i = 1, 2, \ldots, 15$$

in which A_j is the <u>activation</u> of output unit j. This summation formula is one kind of <u>activation function</u> -- a rule which tells the unit what to do with its inputs.

In some models, the activation of a unit depends partly on its previous activation. For example, a model called BSB ("Brain State in a Box") sets up activation functions for its units such that (within pre-set upper and lower bounds)

$$A_j(t+1) = A_j(t) + \text{SUM } x_i w_{ij}$$

in which $A_j(t)$ is j's activation at time t and $A_j(t+1)$ is its activation at t+1 (<u>1</u>).

Other models depart from simply summing the weighted inputs to each unit. One class of models contains "conjuncts" -- two or more input units with a single interconnection to an output unit. The signals from the units in a conjunct are multiplied together before they are multiplied by the weight of the interconnection, thus producing the conjunct's net input to the output unit. If a set of conjuncts feeds into a unit, their net inputs are summed to produce that unit's activation.

<u>Output of a Unit.</u> In general, units interact by sending out signals to other units. The size of a unit's output signal depends on the unit's activation. In a simple model, if the activation exceeds some pre-set "threshold" value, the unit "fires" (i. e., it provides an output); if not, it doesn't. In the simplest case, if a unit fires, its output is 1; if not, its output is 0. The rule which turns a unit's activation into its output is called the unit's <u>transfer function</u>.

In our letter-identifier, an input-layer unit fires when its associated matrix cell is stimulated -- which we assume produces an activation greater than the input-layer unit's threshold. An output-layer unit fires when its activation exceeds its threshold. An output-layer unit's firing means that the unit presents its associated letter as an identification of the input pattern.

The threshold arrangement can be more mathematically complex than is the case in our letter-identifier. Suppose S is the sum of weighted inputs to a unit j whose threshold is T. Some models posit a <u>logistic function</u> such that j's output is

$$o_j = 1 \ / \ (1 + e^{-(S + T)})$$

The arrangement can be still more complex. For example, in a class of neural networks called <u>thermodynamic models,</u> each unit can output 0 or 1, and a stochastic function operating on the unit's inputs determines the probability that its output will be 1 (<u>2</u>).

<u>Training A Network</u>

When we first set up the model, how do we know the values to assign for the weights? Often, we do not. We might start with some arbitrary values and then "teach" the network what to do. For our letter-identifier, we would expose the network to matrix patterns which represent each of the five uppercase letters and note the network's responses (the patterns presented to the model at this point are called the "training set"). Some of the responses may be erroneous at first, and we would have to provide feedback to the network; the result of the feedback is the altering of the weights according to some rule specified by the model.

One well-known procedure for providing feedback and altering the weights is called the <u>delta rule</u> (<u>3</u>). It works in the following way. Suppose, in the initial training trial, our weights are set up so that when our network is presented with the aforementioned vector for "A", it responds that the stimulus could have been either A, B, or E. That is, its output vector is

$$[1,1,0,0,1]$$

instead of the desired target vector

$$[1,0,0,0,0].$$

We subtract the obtained vector from the desired vector by subtracting corresponding elements, yielding the vector of "deltas"

$$[0,-1,0,0,-1].$$

Let us generically label an input vector as I_p (in which "p" stands for "pattern"), an output vector as O_p, and the target vector as T_p. The vector of deltas is then T_p-O_p. Further, let us label an element of an input vector as I_{pi} ("i" denotes an input unit), an element of an output vector as O_{pj}, and an element of the deltas vector as $T_{pj}-O_{pj}$ ("j" denotes an output unit).

According to the delta rule, after a training trial the change (c_{ji}) in the weight of an interconnection between input unit i (like our units 1-15) and output unit j (our units 16-20) depends on the activation I_{pi} of the input unit and the delta $T_{pj}-O_{pj}$ of the output unit:

$$c_{ji} = r(T_{pj} - O_{pj})I_{pi}$$

in which r is the pre-set <u>learning rate</u> of the network, a number which is typically between 0 and 1. The exact value we assign this rate determines how quickly the network converges on its ideal state. It should reflect the degree of "noise" in our training patterns. For example, some of our training "A's" might not be exact "A's". To the extent that they vary from our prototype "A", we would assign a lower value (like .1 or .2) to r; if our inputs are close to our prototype letters, we would assign a higher value (.8 or .9) to r. If we aren't sure how noisy our data are, we'd pick an in-between value. If the value we pick is too low, the network will need many training trials to converge on its target; if the value is too high, it may overshoot the target.

Suppose our learning rate is .9. According to the delta rule, in our example the interconnection between unit 8 and unit 17 (the output unit for "B") should change by

$$c_{17,8} = .9(-1)(1) = -.9$$

as should the other interconnections which caused the wrong output units to fire.

After a network has been trained, its performance is assessed by exposing it to a set of stimuli which resemble the training stimuli but are not identical to them. The goal is for the trained network to classify the test stimuli in the same way that it classifies the training stimuli. The network is judged by its degree of success in achieving this goal, as a network is useful to the extent that it can generalize beyond its training.

Hidden Layers

To this point, our discussion has centered around a network with just an input-layer and an output layer. Most networks which solve problems of practical importance, however, have at least one more layer of units between the input layer and the output layer. Layers between the input and output are said to be <u>hidden</u>. A unit in a hidden layer acts like a unit in any other layer; it takes one or more inputs, and it passes outputs to other units. If we were to add a hidden layer to our model, it would look like Figure 3 (for the sake of clarity, strongly-weighted connections are the only ones shown).

In a trained network, hidden-layer units should correspond to component features of the stimuli. Our letters, for example, can be thought of as being constructed from vertical lines and horizontal lines, each of which is formed from several cells in the input matrix: a left-side vertical line is indicated by units 1, 4, 7, 10, and 13 being on, a crossbar by units 7, 8, and 9, a right-side vertical line by units 3, 6, 9, 12, and 15, a top horizontal line by 1, 2, and 3, and a bottom horizontal line by 13, 14, and 15. A hidden-layer unit with heavily-weighted connections to 7, 8, and 9 would act as a crossbar detector. This unit, in turn, would have strong connections with the output-layer cells which correspond to A, B, and E.

One method of training a network with hidden layers involves an extension of the delta rule called <u>backpropagation</u> (<u>4</u>). This procedure computes weight changes for hidden units by first finding differences between the observed outputs of output-layer units and the desired outputs, and then propagating these differences back to the units which send output values to them. Most units in hidden-layer models operate via complex activation functions (like the logistic function mentioned earlier) which are differentiable and non-decreasing. In these models, the weight change for an output unit j given an input pattern p is

$$c_{pj} = (T_{pj} - O_{pj})f'_j(net_{pj})$$

in which $f'_j(net_{pj})$ is the derivative of j's activation function. The weight change for a hidden unit h is

$$c_{ph} = f'_h(net_{ph})SUM_j c_{pj} w_{hj}$$

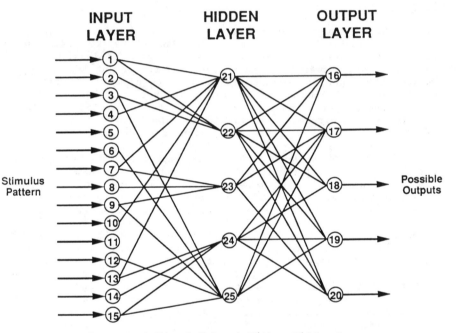

Figure 3. A Neural Network With a Hidden Layer

in which the derivative of h's activation function is multiplied by the weighted sum of changes to units j to which h sends signals. This equation recurses backward through all layers of the network. The overall picture that emerges, then, is a set of signals transmitted forward (from input units through hidden units to output units), and weight adjustments transmitted backward (from output units through hidden units to input units).

Hidden layers are foundational to contemporary work on neural networks. In some models, they allow learning to take place in the absence of training sessions -- i.e., with no feedback from a person. This type of learning is called "unsupervised". One way of accomplishing this is to partition the hidden layer into clusters of mutually inhibitory units (5). Learning takes place via competition among units in a cluster. Each cluster eventually recognizes a different feature of the input pattern. If we add such a partitioned hidden layer to our two-layer letter identifier, we might find that in a 2-unit cluster, one unit learned to respond when a crossbar was present, while the other responded when a crossbar was absent.

Validating A Neural Network. One important aspect of neural network construction is validation. It is not enough to train a network on a set of stimulus patterns and then report the level of accuracy with which the network ultimately comes to identify those patterns. Although network builders also report results of testing stimulus patterns not in the training set, and although these results reflect the reliability of the network, these data do not show the underlying reasons for the decisions that the network makes. In other words, after a network has been trained, what exactly has it learned? A network may be making correct decisions by systematically detecting features of its input stimuli, or by some other process that we cannot interpret as being related to the input in any meaningful way. In the former case, we would tend to believe in the validity of the network as a model of recognition; in the latter case we would not. For any given network, how can we make the distinction?

Analyzing Hidden Unit Activity. Elman and Zipser's (6) study of hidden-layer networks for speech recognition is instructive. These investigators constructed backpropagation networks to classify spoken syllables. Each syllable was a combination of one of three consonants and one of three vowels. The networks consisted of 320 input units, between two and six hidden units (in different conditions), and three or nine output units. The input stimuli were spectral representations of the syllables as spoken by a male speaker. Each spectrum consisted of data for 16 frequency ranges taken over 20 time-segments (each segment represented 3.2 msec); each input unit corresponded to one of the resulting 320 combinations of frequency range and time segment. Each output unit corresponded to one of the syllables to be identified; in one condition, the task was to identify each of the nine syllables, in another, to identify each of the three vowels, and in another, to identify each of the three consonants. One person recorded about 56 repetitions of each consonant-vowel combination, half of which were used for training the network, and half for testing. Starting with randomly-assigned weights, over 100,000 training trials were run, during which a network learned to recognize its training set perfectly. When presented with the test set, the whole-syllable recognizer averaged 16% errors; the vowel-recognizer and the consonant-recognizer averaged 1.5% and 7.9% errors respectively.

Going beyond these performance statistics, Elman and Zipser reasoned that the post-training patterns of activity in the hidden units would provide the best indication of the learning that took place. Using a visual representation that depicted the activity of each hidden unit given each stimulus pattern, they showed that hidden units learned to output a value of 1 for some sound types, and 0 for others.   Thus, every hidden unit became associated with a subset of sound types. These subsets were vowellike or consonantlike, in that a unit was on or off for a particular consonant or vowel. In the case of a four-unit hidden layer for full-syllable recognition, for example, one unit learned not to fire when syllables beginning with "b" were presented, and another learned not to fire when syllables beginning with "g" were presented. Similar results were found for the other two units' responses to vowels. These correlations between stimuli and hidden unit activity are evidence that the network learned to respond to meaningful features of the stimulus set.

Renals and Rohwer (7) found similar results. Their study, an examination of neural networks for recognizing vowels, showed that hidden units learned to respond selectively to members of the stimulus set.

Touretsky and Pomerleau (8) examined hidden-cell activity in a backpropagation network which had been trained to classify computer-generated pictures of road conditions and serve as a navigator for an autonomous land vehicle. Analogous with the phoneme-recognition results, their analysis indicated that the twenty-nine hidden units in the network had learned to respond to component features of the depicted roads.

These studies suggest that studying post-training hidden unit activity is a valuable technique for inferring what a network has learned. Builders of neural networks might do well to consider this technique as standard operating procedure in evaluating their networks' performance.

<u>Scaling The Trained Network's Responses.</u> When cognitive psychologists try to model people's recognition processes for a particular set of stimuli, they study the patterns of errors that people make. For example, when people identify briefly-presented (e. g., 20 msec) lower-case letters of the alphabet they often mistake "q" for "p" and "o" for "c". Mistakes like these have been taken to indicate that people attend to component features of the stimuli; the q-p confusion suggests that each letter's descending straight line has attracted attention, while the o-c confusion suggests the same for letter curvature. These patterns of errors are the foundation for models of the recognition process based on analysis of component features.

A <u>confusion matrix</u> is a convenient way of summarizing error patterns. This is a matrix whose rows are the members of a stimulus set and whose columns are the possible responses to those stimuli (which, ordinarily, are the same objects as the stimuli); each cell of the matrix contains the frequency with which the column's response was given to the row's stimulus (i. e., the "confusability" of the presented stimulus and the resulting response). With perfect recognition, only the cells on the main diagonal are filled, if the potential responses are listed in the same order as the stimulus objects.

Two statistical methods used in behavioral science, multidimensional scaling (9) and hierarchical clustering (10), are used to analyze the error patterns represented in confusion matrices. Together, they can be the basis for a technique which examines the nature of a neural network's decision process.

In contrast with regression-based techniques such as factor analysis, multidimensional scaling is based on the metaphor of "distances" between pairs of objects. That is, two highly confusable objects are characterized as being close to one another in space, two non-confusable objects as being far apart. A confusion matrix is thus viewed as a matrix of interobject distances, and multidimensional scaling procedures map the ordinalities of these distances into spaces of varying numbers of dimensions. Each space is associated with an error term, whose magnitude is inversely related to the number of dimensions. When the number of dimensions is found such that additional dimensions result in no appreciable reduction of the error term, the corresponding space is taken to be the best representation of the confusion matrix. The locations of the stimulus objects in this space enable an experienced analyst to attach labels to the dimensions and to thus make inferences about the nature of the underlying recognition process.

Hierarchical clustering treats highly confusable pairs of objects as belonging to a cluster. When an object from one cluster is confusable with an object from another, the two clusters are joined to form a larger cluster. Iterative applications of this joining process result in a tree-like hierarchy of clusters called a dendogram. Dendograms have been used to delineate the clustering among objects within multidimensional spaces (11), in order to facilitate explanation.

A neural network's responses to its stimuli can be cast into a confusion matrix, so that these techniques could be applied, resulting in insights into a neural network's learning. After high levels of accuracy have been attained as a result of many training trials, a confusion matrix will have little use; the scaling methods could be used throughout training, however, to show how learning evolves. On the other hand, high levels of accuracy on the training set do not necessarily imply high levels on the test set, as Stentiford and Hemmings found in their study of word recognition (12); their post-training test results of 49%-58% accuracy suggest that a confusion matrix and subsequent application of scaling techniques would have been useful in their work. Sawai, Waibel, Miyatake and Shikano (13) reported such a matrix, but did not use it for the application of scaling techniques.

Validation: General Remarks. With the increasing use of multi-layered neural networks for solving real-world problems, users of the technology will become increasingly concerned about how and why a network arrived at a decision. Unfortunately, neural network designers have typically paid little attention to the implications of not validating their networks. In a recent conference on speech recognition (14), for example, over two dozen papers involved neural networks, but the previously-cited paper of Renals and Rowhrer was the

only one which reported hidden unit activity patterns, and the paper
by Sawai et al. was the only one which presented a confusion matrix.

Neural Networks and Expert Systems

Because neural networks and expert systems are both problem-solving
devices, they are often compared. There are several obvious
differences.

First, the knowledge in a neural network is represented not by
explicitly-stated heuristics (as in expert systems), but by the
pattern of numerical values of the interconnections between processing
elements. As these values change, the knowledge representation
changes. Thus, you can't point to a "piece of knowledge" in a neural
net in the same way that you can point to a production rule in a
knowledge base. This leads directly to another distinction: an expert
system can explain its reasoning to a user, while a neural net cannot.

Another difference is the scenario involved in building each type
of device: an expert system comes into being as knowledge is gained
from a human expert through repeated knowledge acquisition sessions;
a neural network, on the other hand, learns to classify patterns,
either by itself or by obtaining feedback from a person (or from a
computer program which generates the patterns in the training set,
presents them, and then checks the neural net's responses to them).

Aside from these differences, several ways suggest themselves for
using these devices together. One frequently-mentioned idea is the
integrated intelligent system, in which a neural network's output (an
identifed pattern) is presented as input to an expert system for
further action.

Another type of synergy may result from neural networks coming
into increasing use as applied problem-solvers. Some people will
become experts at designing the right type of network for a particular
application; this expertise could be captured in an expert system
which, given the characteristics of a problem, could automatically
design a neural network to solve it.

Once a network has been set up, an expert system could be the
basis for an intelligent user-interface between a person and a neural
network. Such an interface could help the person formulate and input
the training set and the test set.

Major Players

At present, many people from industry and from the academic world are
working on neural networks. This section is a brief representative
look at some of this work, not an exhaustive list of all researchers
and groups.

In the academic world, Rumelhart and his colleagues and students
have created an impressive set of models, and they have used these
models to study a wide range of cognitive processes (15). Kohonen (16)
has done some of the pioneering work in systems which exhibit
unsupervised learning. Grossberg (17) has formulated a set of neural
network principles which model phenomena of learning, cognition, motor
control, psychophysiology, and anatomy. Hopfield (18) has shown how
a single-layered network (in which all units are connected to one
another) can store patterns after they have been presented, and
subsequently use these patterns to identify newly-presented ones.

Industry is moving on a number of fronts. Several companies manufacture software shells for constructing neural networks. Others build neural net hardware. Still others provide consulting and training.

Software. Neural network shells usually contain graphics capabilities for illustrating unit activity. In general, their cost is directly related to the number of different types of network models they support.

California Scientific Software's Brainmaker is a low-cost MS/DOS-based program for constructing multi-layer backpropagation networks based on several kinds of transfer functions. It comes with a set of trained networks whose tasks range from shape recognition to text-to-speech conversion.

SAIC's neural network software products are more expensive: ANSim and Shells are environments for implementing frequently-used neural net models; ANSpec is a language for specifying and developing new models. Each package runs on a PC/AT or compatible, or on an AT enhanced with SAIC's Delta accelerator card.

Neuralware Inc.'s NeuralWorks Professional II is a PC-based package for building networks based on a wide range of learning rules and threshold functions. The input data for these networks can be kept in files whose formats are compatible with a number of popular software packages.

Nestor's NDS is a high-end development system for PC/AT's and for Apollo and Sun workstations. Unlike the other neural net shells, NDS is based on a propietary model on which Nestor holds a patent. Nestor claims that networks based on its model can be trained significantly faster than models based on backpropagation.

Hardware. Computers built to work like neural nets are called "parallel processors". A parallel processor uses a large number of small, interconnected processing units rather than a single CPU. Prominent among these is Thinking Machines Corporation's "Connection Machine", which can realize a variety of neural net models. It is programmable in LISP, and is well-suited to database tasks. Hecht-Nielsen produces the ANZA board, a co-processor which allows the PC/AT to emulate a parallel processor.

Consulting. Several firms supply consulting services and reports in the area of neural networks. Perhaps the best known of these is Adaptics, whose president, Maureen Caudill, has written a popular series of introductory articles beginning with (19). Adaptics also provides training for purchasers of SAIC's neural net products. New Science Associates has produced two useful reports oriented toward commercial applications (20,21).

Some Successful Neural Network Applications

Neural network models have been implemented for solving real-world problems in a number of areas. Again, this is not an exhaustive list, but a representative sampling.

SAIC's SNOOPE (22) used backpropagation to learn to detect

plastic explosives in luggage and cargo. It was successfully tested on 40,000 pieces of luggage in June 1988 at the Los Angeles and San Francisco International Airports. It can continuously process 10 bags a minute, and it decides whether or not a bag contains a threat by the time the bag leaves the system.

Nestor has developed the Mortgage Origination Underwriter, which assesses potential borrowers. The input to the system consists of data on the borrower (e. g., credit rating, number of dependents, number of years employed, current income), the mortgage (loan-to-value ratio, type of mortgage, income-to-mortgage ratio), and the property (age, number of units, appraised value). The network can be configured in one of three risk classification modes, depending on the acceptability of an error. It was trained on pool of mortgage applications, and it shows a degree of agreement with a human underwriter.

Sejnowski and Rosenberg's (23) NETtalk, a three-layer backpropagation network, learned to synthesize speech from English text. After training, it could turn text input into phonemic representations which a computer converted to sound. NETtalk was trained on a first grade reading text, and on randomly-ordered words from a dictionary. The input to the network was a "window" of seven characters; NETtalk's output was a phonetic symbol for the center character, the context for which was provided by the other six letters. After each training trial, the window advanced one character position, and NETtalk provided a phonetic symbol for the new center character. NETtalk attained over 90% accuracy after 5 passes through a training set of 1000 words. After the first few training sessions, NETtalk's output sounded like babble, progressed through pseudowords, and began to be understandable by about the tenth pass through the training set.

Speech recognition has proven to be a particularly fruitful field for neural network applications. While the previously-cited study by Elman and Zipser and the study by Renals and Rowhrer showed that networks could learn to classify isolated speech sounds, a number of other investigators have developed networks which learn to recognize whole words. Krause and Hackbarth (24), showed that a network could accurately recognize whole words from a limited German vocabulary. Demichelis, Fissore, Laface, Micca, and Piccolo (25) constructed a network that recognized Italian digit-words, and Sakoe, Isotani, Yoshida, Iso, and Watanabe (26) did the same for Japanese digit-words. In all three studies each hidden unit was connected to a subset of input units, rather to than the entire input layer, as in the phoneme work. This variation shows promise, and may one day lead to networks which recognize large vocabularies and which could be engineered into commercial applications.

A recent survey (27) touched on several other applications: AIWARE has built a system which troubleshoots grinding operations in a factory; Global Holonetics' LIGHTWARE performs quality control on an assembly line; Widrow has developed a neural network which eliminates echoes in telephone lines, and is used in modems and other signaling devices; Carleton University's Neuroplanner is a set of networks which enables a robot to navigate its workspace.

Some Possible Environmental Applications

Decision-makers who deal with problems of the environment typically use large amounts of data from diverse fields in order to make conclusions. Neural networks could help by discovering patterns in the data and making recommendations.

One potential application is the use of neural networks to facilitate decisions about hazardous waste sites. These sites generate a great deal of data, in which patterns are inherent. Sites that once produced batteries, for example, will typically show a great deal of cadmium in the soil; this finding usually leads to a decision about a particular form of remediation. A network's input layer could represent characteristics of hazardous sites (such as type of site, volume of contamination, type of contaminants, contaminated media, etc.), and its output units could correspond to possible decisions regarding methods of cleanup. Such a network could be trained and tested on RODs (Records of Decision) to establish the appropriate relationships and assess the network's accuracy.

In the same vein, a network could be used for estimates of level-of-effort and financing for the RI/FS (Remedial Investigation/Feasibility Study), the initial stage of hazardous waste cleanup. Again, using site characteristics as input and historical data for training and testing, a network could learn to arrive at reliable, consistent estimates, thus facilitating what has usually been a costly and time-consuming budgeting process.

Two other pattern-recognition based applications suggest themselves: (a) analysis of soil and liquid samples, and (b) geophysical exploration for groundwater.

Soil and Liquid Analysis. One technique for analyzing soils and liquids combines gas chromatography with mass spectrometry. In this procedure, gas chromatography is used to separate and ionize the components of a soil or liquid mixture which has been converted to a gas. The ionized components are then passed to and through a mass spectrometer, whose mass analyzer sorts the ions into beams of the same mass-to-charge ratio. The spectrometer's detection system detects these mass-analyzed ions either photographically or electronically, and the spectrometer's recorder ultimately produces a frequency distribution of mass-to-charge ratios for each component. This distribution, called a "fingerprint", is often identified by matching it against a computerized library of typical fingerprints for substances. Neural networks represent an alternative way of identifying these fingerprints. The input layer would represent several ranges of frequencies for each of a wide range of mass-to-charge ratios. The output layer would have one unit for each possible substance to be identified.

Exploration for Groundwater. Surface and subsurface methods for geophysical exploration for groundwater result in patterns of data whose interpretation requires training and experience. One surface method, seismic refraction, takes advantage of the fact that an elastic wave's velocity through earth material varies with the density of the material. When an elastic wave crosses over a geologic boundary between two formations with different elastic properties, the wave's path is refracted. In this type of geophysical exploration, elastic waves are generated by a small explosion (or sometimes by the blow of a hammer) at the surface. A set of receivers called geophones is situated in a line radiating outward from the origination point of the waves. The waves follow three types of paths to the geophones -- direct (along the surface), refracted, and reflected. The time for a wave to reach a geophone will depend on the path it takes and the density of the material. The time-distance relationships which form the basis of the data analysis are often complicated by the presence of several distinct layers of sediment. To facilitate the analysis, a neural network could be taught to classify typical patterns of wave arrivals. The input layer would represent a set of wave amplitudes over time and distance; an input unit would fire only if its amplitude-time-distance combination was represented in the data.

Another surface exploration method is based on electrical resistivity of a geological formation. In a soil or rock that has been saturated with fluid, the resistivity depends mainly on the porosity and the density of the material and on the salinity of the saturating fluid. In an electrical resistivity survey, two current electrodes pass an electric current into the ground, and the potential drop is measured across a pair of potential electrodes; the spacing between the current electrodes determines the depth of penetration. The resistivity is calculated from the measured potential drop, the applied current, and the electrode spacing; as resistivity values change (either with increasing depth in one location, or at one depth over many locations), they indicate change in subsurface conditions. Resistivity is plotted against electrode spacing, and the plot is compared against published plots to provide stratigraphic interpretation. A neural network could be trained on the published resistivity plots, and then be used to interpret the obtained plots from resistivity surveys. The input layer would represent a set of resistivity ranges for different electrode spacings, and the output units would correspond to the possible stratigraphic interpretations.

Many subsurface methods are based on borehole geophysics -- a set of techniques in which a sensing device is lowered into a hole to gather data which are then interpreted in terms of the characteristics of the geologic formations and the fluids they contain. One frequently-used sensing device utilizes an electrode dropped into a borehole, one at the surface, and a source of current. Two data-records are gathered -- (a) the potential difference (vs. depth) between the borehole electrode and the surface electrode with the current source turned off, and (b) resistivity vs. depth for a given current strength. The spikes contained in these records constitute an ideal type of data for interpretation by a neural network. In either case, the input layer would represent ranges of spike amplitudes over a range of depths. The network could be trained and tested on published data, and used to interpret the spike patterns in the obtained records.

Environmental Applications: Conclusion. Neural networks are not without their deficiencies. To attain suitably high levels of accuracy, they require a great deal of training and computational resources. Also, because current interest in neural networks is relatively new, the optimum neural net architecture is not yet known for every type of problem; indeed, finding the ideal architectures for various problem-classes is a continuing research area (28).

Nevertheless, as organizations concerned with the environment come to rely on increasingly large data sets (and on the automation of the management of these data sets), they are likely to turn to neural networks to help use these data to reach decisions.

Literature Cited

1. Anderson, J. A. In Basic Processes in Reading, Perception and Comprehension; LaBerge, D. and Samuels, S. J., Eds.; Erlbaum: Hillsdale, NJ, 1977, pp 27-90.

2. Rumelhart, D. E.; Hinton, G. E.; McClelland, J. L. In Parallel Distributed Processing; Rumelhart, D. E. and McClelland, J. L., Eds.; MIT Press: Cambridge, MA, 1987; Vol. 1, pp 45-76.

3. Widrow, G.; Hoff, M. E.; Western Electronic Show and Convention, Convention Record, 1960, Pt. 4, pp 96-104.

4. Jones, W. P.; Hoskins, J. Byte 1987, 12 (10), pp 155-162.

5. Rumelhart, D. E.; Zipser, D. In Parallel Distributed Processing; Rumelhart, D. E. and McClelland, J. L., Eds.; MIT Press: Cambridge, MA, 1987; Vol. 1, pp 151-193.

6. Elman, J. L.; Zipser, D. J. Acoust. Soc. Am. 1988, 83 (4), pp 1615-1626.

7. Renals, S.; Rohwer, R. Proc Int. Conf. on Acoustics, Speech, and Signal Processing, 1989, pp 413-416.

8. Touretsky, D.; Pomerleau, D. Byte 1989, 14(8), pp 227-233.

9. Kruskal, J. B. Psychometrika 1964 29, pp 1-27.

10. Johnson, S. C. Psychometrika 1967 32, pp 241-254.

11. Schmuller, J. Brain and Language 1979 8, pp 263-274.

12. Stentiford, F.; Hemmings, R. Proc Int. Conf. on Acoustics, Speech, and Signal Processing, 1989, pp 310-313.

13. Sawai, H.; Waibel, A.; Miyatake, M.; Shikano, K. Proc Int Conf on Acoustics, Speech, and Signal Processing, 1989, pp 25-28.

14. Proc. Int. Conf. on Acoustics, Speech, and Signal Processing, IEEE: Glasgow, Scotland, 1989.

15. Rumelhart D. E.; McClelland, J. L. Parallel Distributed Processing; MIT Press: Cambridge, MA, 1986; Vols. 1 and 2.

16. Kohonen, T. Self Organization and Associative Memory; Springer-Verlag: New York, 1988.

17. Grossberg, S. Studies of Mind and Brain; Reidel: Dodrecht, Holland, 1982.

18. Hopfield, J. J. Proc. Natl. Acad. Sci., USA, 1982, pp. 2554-2558.

19. Caudill, M. AI Expert 1987, 2 (12), pp. 46-52.

20. Industry Report: Artificial Intelligence Research; New Science Associates: South Norwalk, CT, 1988(Spring).

21. Research Highlights: Artificial Intelligence Research; New Science Associates: South Norwalk, CT, 1989(July).

22. Obermeier, K. K.; Barron, J. J. Byte 1989, 14 (8), pp. 217-224.

23. Sejnowski, T. J.; Rosenberg, C. R. JHU/EECS-86/01, School of Electrical Engineering and Computer Science, Johns Hopkins University, 1986.

24. Krause, A.; Hackbarth, H. Proc. Int. Conf. on Acoustics, Speech, and Signal Processing, 1989, pp 21-24.

25. Dimichelis, E.; Fissore, L.; LaFace, P.; Micca, G.; Piccolo, E. Proc. Int. Conf. on Acoustics, Speech, and Signal Processing, 1989, pp 314-317.

26. Sakoe, H.; Isotani, R.; Yoshida, K.; Iso, K. I.; Watanabe, T. Proc. Int. Conf. on Acoustics, Speech, and Signal Processing, 1989, pp 29-33.

27. Thompson, D.; Bailey, D., Feinstein, J. PC AI 1989, 3(2), pp. 56-58.

28. Hinton, G. E. CMU-CS-87-115, Computer Science Department, Carnegie-Mellon University, 1987.

RECEIVED April 23, 1990

Chapter 5

Expert Systems To Support Environmental Sampling, Analysis, and Data Validation

Ramon A. Olivero[1] and David W. Bottrell[2]

[1]Environmental Programs Office, Lockheed Engineering and Sciences Company, 1050 East Flamingo Road, Las Vegas, NV 89119
[2]Environmental Monitoring Systems Laboratory—Las Vegas, U.S. Environmental Protection Agency, P.O. Box 93748, Las Vegas, NV 89193-3478

Expert systems are being developed to address the decision-making needs for data generation activities (i.e., sampling, analysis, and data validation) at the U.S. Environmental Protection Agency. This paper describes the Environmental Sampling Expert System, the Smart Method Index, and Computer-Aided Data Review and Evaluation, among other systems under development at the Agency's Environmental Monitoring Systems Laboratory in Las Vegas, Nevada, and discusses their design, operation, and impact on environmental investigations. Appropriate data quality is fundamental to environmental decision making, monitoring, and remediation. Requirements for standardization and documentation and the need for rapid response from personnel with different levels of training make the application of expert system technology a promising approach for the Agency. Improvement in quality and consistency of environmental data through the application of expert systems in the government and private sectors is expected to translate into lower costs, from both economic and social perspectives.

The U.S. Congress has tasked the U.S. Environmental Protection Agency (EPA) with directing and overseeing the effort to control and remediate pollution nationwide. The Superfund program was specifically established to assess and remediate existing hazardous waste sites. Superfund technical evaluations help set cleanup priorities for sites according to the risk posed to human and ecological health.

Investigations at hazardous waste sites involve planning, management, data collection, risk assessment, technology selection, and engineering design and construction on a very large scale. The decision-making responsibilities are numerous and varied. Often

0097–6156/90/0431–0069$06.00/0
© 1990 American Chemical Society

they are beyond what can be efficiently or reliably performed by the available personnel. This may be due to the breadth of experience required or to the amount of available data that is relevant to a particular decision.

Computers are being increasingly applied by EPA to help expedite the Agency's work. In addition to task-automation computer programs, systems that use artificial intelligence techniques are being developed to serve as "smart" advisors for decision makers at many different levels. The EPA Environmental Monitoring Systems Laboratory in Las Vegas, Nevada, is developing expert systems to increase the accuracy, timeliness, and cost effectiveness of field sampling, chemical analysis, and analytical data validation within the Superfund program.

EPA Decision-Making Needs

Decisions made by EPA staff and contractors cover a wide range, depending on the nature of the problem and the stage of project activity. A preliminary study using risk assessment techniques may be concerned with establishing the existence and extent of an environmental hazard. Typically, the data requirements are the concentrations of pollutants in a specific environment (e.g., levels of pesticides in a subsurface water source). At more advanced stages, decisions must be made about the need for remedial actions or pollutant-generation controls and the selection of appropriate technologies for implementation. The process is monitored throughout to establish progress, verify attainment of objectives, or assure continued compliance. All of these decisions require quantitative and qualitative information of known quality and appropriate for the intended use. The efficient interpretation and application of adequate data is the cornerstone for sound decision-making in all areas of the EPA mission.

EPA utilizes personnel from many areas of expertise. For example, a particular project may involve people with backgrounds in environmental engineering, health and safety, chemistry, earth sciences, statistics, construction engineering, management, and law, among others. Many of the specific skills needed to plan and execute the variety of activities involved in environmental investigations are learned through experience and specialized training, rather than through formal education. The body of specialized knowledge developed in the recent past by the environmental community is not integrated. Technologies develop faster than documentation and, in many cases, the knowledge is more empirical than theoretical.

Each member of an environmental team has a very specific mission. The success of the project depends on the overall coordination of the individual elements. Communication and cross-training are critical for the effective and efficient accomplishment of the EPA mission. Unfortunately, the demand for trained and experienced environmental professionals far outweighs the current availability.

EPA has undertaken the development of computerized information systems, decision support systems, and "smart" advisors to provide access to the specialized knowledge of experts and confront understaffing and high personnel turnover rate. Advisory software is commonly termed "expert systems" or "knowledge systems." To achieve a high level of performance for a particular task, this type of computer program incorporates the knowledge and simulates the decision-making processes of human experts. Expert systems have the potential for increasing the accuracy, timeliness, and consistency of decisions.

Decision-making for environmental work requires combining information obtained from environmental data and standard procedures with judgement. It should be based on the best data and information available, follow existing EPA regulations, and be carried out by skilled personnel. Expert systems can guide the user to relevant data bases and regulations (and help in their interpretation and application), as well as offer specific advice based on the experience of human experts. A key area for the application of expert systems is quality assurance for data that are to be used as input for decision making in the various aspects of environmental work. These systems not only have a beneficial impact on the quality, timeliness, and cost-effectiveness of the data-generation process itself, but also have a positive effect on the decision-making and action phases of projects.

Environmental Data-Generation Process

EPA has developed a standardized procedure for generation of data for environmental decisions. The data quality objective (DQO) procedure establishes a sequence of ordered steps to assure that the data generated is of known quality and appropriate for the intended use (1). Specific DQO guidelines have been developed for Superfund-related work (2).

The overall process of generating environmental information involves field sampling, chemical analysis of the collected samples, validation of the data collected, and evaluation of the quality and useability of the data based on the pre-established DQOs.

DQOs include statements in terms of precision, accuracy, detectability, representativeness, comparability, and completeness of the analytical data. These quantitative parameters are used to select appropriate sampling and analysis techniques. After sample collection and analysis, data quality is assessed to establish the degree of attainment of the DQOs. To control, monitor, and correct the process, quality assurance and quality control (QA/QC) procedures are implemented throughout.

Scientists at the Las Vegas laboratory are developing expert systems to support various key aspects of this data-generation process. The aspects currently being addressed include selection of sampling techniques, selection of chemical analytical methods, evaluation of analytical laboratory performance, and data validation. Figure 1 depicts the phases of the data-generation process and systems being developed at the Las Vegas laboratory to address them. The EPA Quality Assurance Management Staff in Washington, D.C., is developing an expert system to assist in establishing DQOs.

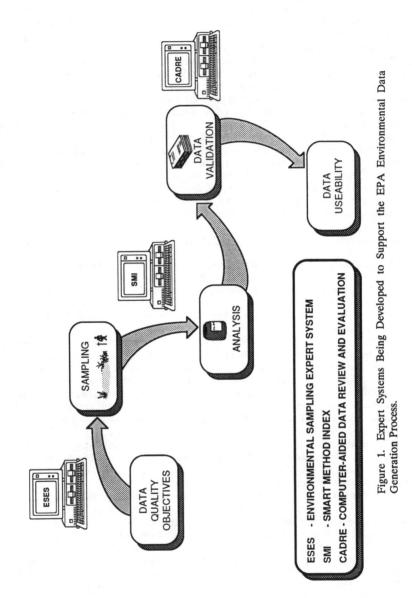

Figure 1. Expert Systems Being Developed to Support the EPA Environmental Data Generation Process.

Environmental Sampling Expert System

Sampling and analysis constitute the two major components of the measurement phase. To a great extent, their combined individual errors determine the overall measurement error. Development, improvement, and characterization activities for chemical analysis technology have received more attention than those for field sampling technology. Significant progress has been made in laboratory and field analysis techniques, but sampling remains difficult to control, largely due to its complexity and diversity. Consequently, the field sampling step has the potential to introduce unaccounted uncertainty in the data.

The objective of the Environmental Sampling Expert System (ESES) is to consolidate knowledge of alternative techniques in order to plan sampling activities at hazardous waste sites in an efficient, consistent, and coordinated fashion. The system is built in modules that address various aspects of sample collection. Input is based on project DQOs and site characteristics. The user is expected to be familiar with the DQO process and provide necessary background information on the site. ESES makes extensive use of hypertext techniques (3). Hypertext is a method to present information in a computer. Each portion of text presented on the computer screen may contain highlighted terms which can be selected for further explanation. The explanation will appear in a separate "window" on the screen and the user, can return to the original text when finished with the window. Each explanation window may contain more hypertext terms, which can be selected at the user's command, forming a chain of concepts. This feature allows for various levels of on-screen information according to the user's level of background in the subject matter. Novice users will make extensive use of hypertext and find the system self-explanatory, while more experienced users will not be forced to read familiar information. The use of hypertext gives ESES value as a training tool and also makes it appropriate for use by staff members with wide-ranging backgrounds and levels of expertise.

The ESES prototype was developed with the KnowledgePro Version 1.4 "knowledge processing" software package (Knowledge Garden, Inc., Nassau, New York) (4). This software has built-in facilities for hypertext and hypergraphics manipulation, as well as extensive support for user interface development. A backward-chaining inference engine permits the implementation of decision rules. In KnowledgePro data are kept in lists and several list-manipulation procedures, similar to LISP programming, are provided. External links to memory-resident Pascal code can also be implemented. ESES takes advantage of this useful feature to implement statistical routines. However, the interface scheme is not very efficient, and thus forces the programmer to limit the use of external routines to the most essential tasks.

The ESES expert system provides an explanation facility which justifies the recommendations given to the user (this is the HOW feature, also frequently referred to as the WHY feature in expert system literature). A comprehensive report of the session can be printed for future reference; which includes the recommendations given, HOW explanations, a profile of the problem described by the

user, and a copy of the hypertext explanations requested.
Additionally, sessions can be saved to resume consultation at a
later time.
    As ESES has grown in size and complexity, the IBM-PC platform
and KnowledgePro 1.4 software have become less capable of adequate
performance. To obtain acceptable performance, the system must be
used on an 80286-based machine running at least at 12 MHz clock
speed. Therefore, an 80386-based microcomputer is a better
environment for this system. EPA needs expert systems that run on
its extensive installed base of IBM-PC compatible microcomputers.
This requirement is the major hardware constraint for any system
developed for Agency-wide distribution. Furthermore, enhancements
such as extended memory; new, faster microprocessors; massive
storage systems; and even special pointing devices are not
available, at this time, throughout the Agency.
    Given these hardware limitations, the system improvement effort
has concentrated on the software component. The use of a faster
development software product with better input and output
capabilities will allow for greater modularization and more storage
of temporary information on disk, thus easing main memory usage and
facilitating the interfacing to external routines. This kind of
software is expected to significantly increase performance. The
next generation of KnowledgePro software promises to meet these
requirements and is being currently tested, in its beta version, by
the Las Vegas laboratory.
    Two versions of ESES are currently under development, the soil
metals application (ESES-SM) and the ground-water application (ESES-
GW). Knowledge engineering for the system (the process of acquiring
and organizing the knowledge and decision rules) is done by
iterative interviewing of recognized experts in the areas of field
sampling, soil science, chemistry, hydrogeology, statistics, and
quality assurance. ESES-SM assists in designing a sampling plan
for determining the extent of metal pollution in soil (5). It
provides advice on appropriate statistical designs, QA/QC
procedures, sampling techniques and tools, sample handling, budget
requirements, personnel safety, and documentation. ESES-GW has an
extended analyte coverage that includes organic contaminants in
addition to metal contaminants (6). The current ESES-GW prototype
helps decide what types of ground-water sampling pumps and devices
are appropriate to use under given site conditions. Advice is also
given on proper sample handling, field determinations, QA/QC
procedures, personnel safety measures, and documentation. Planned
areas of expansion include surface-water sampling and soil organic
contaminant sampling applications. ESES characteristics, as well
as those of the next two systems described, are summarized in Table
1.

Smart Method Index

A significant number of sample analyses for environmental monitoring
are performed under legislative mandate. EPA is required to monitor
waste sites and the quality of the environment in general.
Pollutant generators, handlers, and disposers are required by
regulation to monitor their operations. In fact, legislation is a

Table 1. Characteristics of Quality Assurance Expert Systems Described

SYSTEM	DOMAIN	LANGUAGE/ TOOL	PLATFORM	AI TECHNIQUES	FEATURES
ESES	Sampling Plan Preparation/ Review	KnowledgePro Pascal	IBM-PC 286/386	Rule-driven Inference	Hypertext Hypergraphics Modularization
SMI	Analytical Method Selection	SAS Prolog	Mainframe/Mini IBM-PC	Natural Language Understanding	Centralization Easy Query
CADRE	Data Validation	Pascal	IBM-PC	Object-Oriented Programming	User-customizable Data Security

catalyst for analytical methods development and improvement. At the same time, the state of currently available capabilities is a limiting factor for the levels of regulatory monitoring required.

Information on availability, characteristics, applicability, and performance of analytical methods is often inconsistent and not readily accessible. Method information is fragmented throughout EPA program documentation and between other agencies and institutions. A needs assessment study, performed to support the EPA Expert System Initiative, indicated a high priority for the development of an intelligent method index. Some benefits expected from this system are faster and better method selection, use of more appropriate DQOs, reduction of duplication of effort, and the identification of areas for further research.

The Smart Method Index (SMI) system is designed to sort options according to information need, analytes, matrix, performance, applicable regulations, and other criteria and to retrieve for the user analytical methods applicable to the problem (7). The SMI user base may include analytical chemists, researchers, project managers, industry personnel, legislative staff, and concerned citizens. Essential design requirements are comprehensiveness, accessibility, and relative ease of use. The current development focuses on a hybrid implementation. A centralized data base manager in a mainframe computer is remotely accessed via a microcomputer-based smart user interface. Natural language techniques are being investigated to provide the user with an English-like query facility. This microcomputer-based component is being developed in Prolog language. It has the function of translating the user's query to the code that will cause the data base management system to retrieve the desired data. Widely experienced analytical chemists are involved in the specification of the basic English queries to be supported by the system. The contents of already existing method data bases in EPA are being integrated with SMI, thus avoiding duplication of costs on data gathering. An added benefit for the users is the ability to query a number of different data bases through a single interface. The need of a large number of users to access this large data base makes centralization a must because the logistics of distribution would be practically insurmountable. This approach is also consistent with the necessity for "instant updating."

A secondary objective of this project is to explore the development of expert systems in a mainframe platform in the EPA network. Intelligent system components prototyped in the microcomputer could be implemented on a mainframe computer. This is an alternative to the expense of mainframe development that has deterred the implementation of expert systems for this platform at EPA. The relatively high cost of expert system development tools for mainframe computers could be obviated by transferring expert system shells developed by other Government agencies to the EPA. Some of the choices are CLIPS, developed by the National Aeronautics and Space Administration, and LES (Lockheed Expert System), developed by Lockheed within Department of Defense projects.

Computer-Aided Data Review and Evaluation

Most Superfund data are obtained through the EPA Contract Laboratory Program (CLP). Data are produced by approximately one hundred independent laboratories and utilized by the ten EPA Regions. The results of the analyses are routinely reviewed and validated against standard criteria to assure that they are of known quality, applicable for their intended use, and legally admissible (8,9).

Data review and evaluation has been determined to be rate-limiting in the data generation process due to its labor-intensive nature. EPA Regions accumulate a backlog of several thousand samples for review. In some instances the review is not as thorough as intended due to the tradeoff of completeness and accuracy for timeliness.

The Computer-Aided Data Review and Evaluation (CADRE) system assists in the validation and reporting of information by automating most of the QA/QC checks for electronically delivered data (10). Figure 2 shows the integration of CADRE in the general CLP data flow, including the mainframe-based CLP Analytical Results Data Base (CARD). CADRE follows general data validation rules determined by the EPA Analytical Operations Branch and Regional offices, complemented with the judgement of CLP methods quality assurance experts. It can be customized by each Region to accommodate local conditions, special project needs, and non-CLP data. In this decentralized environment, capabilities for user customization are an essential requirement for the system utilization. The microcomputer platform, selected for CADRE development and delivery, is ideal for this type of application.

Object-oriented programming techniques (as an extension to the Pascal programming language) are being applied in CADRE because of the advantage of re-using pieces of code in different versions with minimal reprogramming. This programming technique makes pieces of code and data "smart;" they act as entities with the details of their inner workings hidden from the outside. Code sections pass "messages" to each other and respond to the other code section messages to make the overall assembly work. The CADRE CLP ORGANIC version automates the process of validation of organic analysis data (volatile, semivolatile, and pesticide compounds). CADRE CLP INORGANIC and Quick Turnaround CADRE, are under development to automate validation of inorganic data and Quick Turnaround Methods data, respectively.

Other Developments

Other artificial intelligence techniques that are being investigated at the Las Vegas laboratory include the application of adaptive neural networks to pattern recognition of mass spectra of organic compounds. This application could serve as a complement to the current library search method for identifying unknown pollutants by gas chromatography/mass spectrometry analysis. The library search method may misidentify compounds due to shortcomings in the search algorithm or to sample complexity (e.g., coelution of analytes). The spectra in the present NIST/EPA library include over 50,000 entries. An effective pattern recognition method could provide at

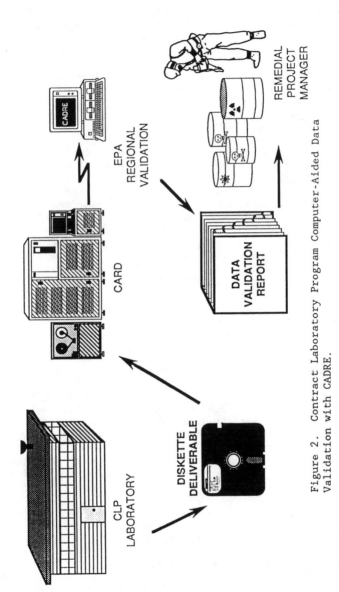

Figure 2.   Contract Laboratory Program Computer-Aided Data Validation with CADRE.

least a chemical class identification, giving some information as to its potential for environmental impact.

Laboratory performance evaluation systems are being designed and developed (11) to help monitor and qualify the performance of laboratories in the CLP and to assist in the selection of appropriate laboratories for contracting analytical work of a specific nature. Both the evaluation of laboratories for participation in an environmental investigation and monitoring of the facility during operation are significant aspects of the data collection process. The development of performance-based analyte- and matrix-specific quality control components of analytical methods is essential to the identification of confidence limits for both detection and quantification of contaminants. Current knowledge (e.g., instrument performance criteria) and areas of research (e.g., surrogate/analyte correlation) can be integrated in an expert system appropriate for the evaluation of laboratory, method, and program performance. A system utilizing similar components for the evaluation of initial performance on a qualification sample is appropriate as a primary consideration in the selection of a contractor facility. Other considerations that may be appropriate include potential conflict of interest, cost, proximity to site, and analytical capacity.

Summary and Discussion

Developers of expert systems for the EPA must recognize important Agency considerations, such as the varying nature of the problems addressed, a staff with diverse experience, decentralization of tasks, computer resource constraints, geographical distance, and high personnel turnover.

During the design phase, close attention should be paid to selecting the appropriate delivery environment for each system's target user base. Expert systems present particular difficulty because of the high demand on computer resources associated with this type of software. It is difficult to fit expert systems into computer hardware that must also be used for routine tasks. At the same time, the introduction of specialized platforms for delivery of expert systems would be costly and would present an acceptance problem which might prevent integration of expert systems into the mainstream EPA computing environment together with other data handling and analysis software.

An effort has been made to provide compatible, if not standard, user interfaces across different systems (e.g., menu-driven interfaces). This is more difficult when expert system development shells, with their own specifics, are used for development. Many aspects of expert systems, including operation, documentation, and system life cycle management, do not lend themselves very well to a high level of standardization. An important consideration in standardizing development is compliance with EPA system life cycle management guidelines (12). These guidelines are stated in general terms and assure that appropriate consideration has been given to cost-benefit concerns, to project relevance, and to funding of development and maintenance. Their primary purpose is to ensure that the system will adequately perform the function for which it was intended. With this in mind, the general system life cycle management requirements must be customized or adjusted, to a degree, for each system to achieve its goals, while at the same time avoiding increasing the development cost beyond the point of diminishing return.

The nature of expert systems and their development process presents particular difficulties for the application of the traditional method of validating software against comprehensive pre-specifications, since the final level of operation of an expert system is difficult to pre-determine. Testing by measuring the degree of comparability of the system's performance to the performance of human domain experts, by using a representative set of problem cases, is more practical and more in tune with the definition of expert systems.

Expert systems are a promising approach to improving environmental data generation in terms of quality, timeliness, and cost-effectiveness. Improvements in this crucial area of activity will benefit not only the data collection programs, but also will increase the efficiency and effectiveness of remedial programs.

Notice

The information in this document has been funded wholly or in part by the United States Environmental Protection Agency under contract number 68-03-3249 to Lockheed Engineering & Sciences Company. It has been subjected to Agency review and approved for publication. Mention of trade names or commercial products does not constitute endorsement or recommendation for use.

Literature Cited

1. Development of Data Quality Objectives (internal document), U.S. Environmental Protection Agency, Quality Assurance Management Staff, Washington, DC, 1986.
2. Data Quality Objectives for Remedial Response Activities-- Development Process, EPA/540/G-87/003, U.S. Environmental Protection Agency, Office of Solid Waste and Emergency Response, Washington, DC, 1987.
3. Olivero, R. A.; York, K. R.; Homsher, M. T.; Cabble, K. J., A Hypertext-Based System for Planning of Environmental Sampling Projects, In Proc. 4th Annual Lockheed Artificial Intelligence Symp., Calabasas, California, 1989, pp 4-81 - 4-91.
4. KnowledgePro User Manual, Version 1.4, Knowledge Garden, Inc., Nassau, New York, 1988.
5. Olivero, R. A.; Cameron, R. E.; Cabble, K. J.; Homsher, M. T.; Stapanian, M. A.; Brown, K. W., Environmental Field Sampling Expert System--Development of a Soil Sampling Advisor, In Proc. 1st Intl. Symp. on Field Screening Methods for Hazardous Waste Site Investigations, Las Vegas, Nevada, 1988, pp 325-339.
6. Cameron, R. E.; Olivero, R. A.; Cabble, K. J.; Carlsen, C.; Teubner, M. D.; Bottrell, D. W.; Homsher, M. T., An Expert System Approach for Selection of Sampling Methods for Ground-Water Contamination at Hazardous Sites, In Proc. of the Intl.

Conf. Chemistry for the Protection of the Environment, Lublin, Poland, 1989.

7. Olivero, R. A.; Boyd, J. L.; Bottrell, D. W.; Homsher, M. T., A Smart System for Selecting Analytical Methods for Environmental Analysis--Concept and Design, In Proc. 5th Annual Waste Testing and Quality Assurance Symp., U.S. Environmental Protection Agency, Washington, DC, 1989, pp I-217 - I-226.

8. Hellmann, M. A.; Cheatham, R. A., Data Validation--Its Importance in Health Risk Assessments, Environ. Sci. Technol., 1989, 23(6), pp 638-640.

9. Fairless, B. J.; Bates, D. I., Estimating the Quality of Environmental Data, Pollution Engineering, 1989, pp 108-111.

10. Shumann, C. R.; Olivero, R. A.; Homsher, M. T.; Petty, J. D., Automation of Regional Data Validation, In Symposium on Waste Testing and Quality Assurance: Third Volume, ASTM STP 1075, D. Friedman, Ed.; American Society for Testing and Materials, Philadelphia, 1989.

11. Homsher, M. T.; Olivero, R. A.; Robertson, G. L.; Moore, J. M.; Fisk, J. F., Development of an Expert System for the Analysis of Laboratory Performance Evaluation Data, In Proc. 3rd Annual Lockheed Artificial Intelligence and Strategic Computing Symp., Houston, Texas, 1987.

12. System Life Cycle Management Guidance Practice Paper for Expert Systems, U.S. Environmental Protection Agency, Office of Solid Waste and Emergency Response, Washington, DC, 1988.

RECEIVED March 12, 1990

Chapter 6

An Intelligent Quality Assurance Planner for Environmental Data

Functional Requirements

Nitin Pandit[1], John Mateo[2], and William Coakley[3]

[1]Roy F. Weston, Inc., 955 L'Enfant Plaza, S.W., Sixth Floor, Washington, DC 20024
[2]Roy F. Weston, Inc., Response Engineering and Analytic Contract Project, Woodbridge Ave., Edison, NJ 08837
[3]Environmental Response Team, U.S. Environmental Protection Agency, Woodbridge Ave., Edison, NJ 08837

To clean up hazardous waste sites under Superfund, the Environmental Protection Agency's Removal Program has developed guidance to assist their On-Scene Coordinators specify QA requirements for the collection and analysis of environmental samples to meet the data use objectives in a consistent manner. To use the guidance to its maximum potential, one requires significant expertise in multiple disciplines. IQAP is proposed to be an intelligent computer program that encodes the expertise of experts in developing an acceptable QA plan. The key functional requirement is to emulate how experts correlate their understanding of the site conditions with the available tools in the context of the data use activity to provide an efficient data transmission system and expert systems for data validation and sampling completes a fully functional framework of a total system.

The Environmental Protection Agency's (EPA) Quality Assurance Management Staff (QAMS) has developed guidance (1) to assist EPA's program offices in preparing a program-specific Quality Assurance (QA) plan. To meet these specifications, guidance for region-specific QA plans was incorporated by QAMS in (1). Similarly, the Removal Program has adopted the guidance in (2) to prepare generic QA Project Plans, and the Office of Solid Waste and Emergency Response (OSWER) Directive 9360.4-01 (3) to prepare site-specific Sampling Quality Assurance and Quality Control (QA/QC) Plans.

For a site activity under the purview of the Removal Program, the On-Scene Coordinator (OSC) has the responsibility of preparing the site-specific Sampling QA/QC Plan. The objective of the Sampling QA/QC Plan is to ensure that field sampling efforts and analytical services will provide data of known quality [The quality of data are known when all components associated with their derivation are thoroughly documented, with such documentation being verifiable and defensible.]. The various activities of the OSC in the life-cycle of the data, from the preparation of the Sampling QA/QC Plan through the use of the validated analytical results are complex and extensive, requiring significant expertise. The Environmental Response Team (ERT) has made significant progress in supporting the OSC's activities through automated systems.

This paper describes one of these systems called IQAP, an Intelligent Quality Assurance Planner. The needs of the OSC are discussed first and how IQAP is designed to address a specific need in the integrated OSC support system. The next section describes the system inputs and outputs, respectively. Finally, there is a discussion of how Artificial Intelligence (AI) techniques will be used to embed the expertise of experienced OSCs, field and lab personnel, and QA personnel in the development IQAP.

User Needs

This section describes two activities of the OSC which can be supported by automated systems. They are:

o   selection of a QA/QC Objective for a field data collection activity

o   using the QA/QC Objective to make concrete specifications for sampling and for the analytical methodology.

These activities are performed sequentially. Consequently, a single system can be developed to assist the OSC. However, there are significant differences in the knowledge and expertise needed to perform these two activities. In view of this, their development was taken up separately as described in the sections below.

Selection of the QA/QC Objective. The OSWER Directive 9360.4-01 describes three distinct categories (or levels) of QA/QC Objectives, referred to as QA1, QA2, and QA3. The OSWER Directive notes that before any sampling activity is conducted, a determination must be made regarding the intended use of the data and the OSC must consider the relative importance of the data before determining the QA/QC Objective. This is useful to remember, because it implies that a carefully specified QA/QC Objective inherently embodies data usability criteria.

Table 1 presents the rationale, characteristics, potential applications, and primary QA/QC requirements for each of the three QA/QC Objectives. In Table 1 those sampling activities requiring a QA1 Objective are shown as associated with minimal QA requirements and those requiring QA3 Objective are associated with rigorous QA requirements. In a broad sense, the QA3 Objective imposes QA requirements similar to those specified by EPA's Contract Laboratory Program (CLP). The selection of the QA/QC Objective is a process requiring expertise; consequently, ERT has started an effort to embed the expertise in an expert system. An expert system is a computer program that embeds the judgments and expertise of expert(s) using AI techniques.

Table 1. QA/QC Objectives

	QA1	QA2	QA3
Rationale	Allow for the collection of the greatest amount of data with the least expenditure of time and money, using rapid, non-rigorous methods of analysis to make quick, preliminary assessments of types and levels of pollutants.	Provide a level of confidence for a select portion of preliminary data, focus on specific pollutants and specific levels of concentration quickly, by using field screening methods and verifying 10% by more rigorous analytical methods and quality assurance.	Provide a level of confidence, using rigorous methods of analysis, for a select group of "critical samples" (samples for which the data are considered essential in making a decision)
Characteristics	No QA data collected. Analyte or non-analyte specific (Can be chemical class specific) Non-qualitative to semi-qualitative, non-definitive (unconfirmed) identification. Non-definitive quantitation; no confidence limits.	Analyte specific. Verification of preliminary screening results by (Choose ONE): • definitive identification of at least 10% of samples with organic analytes. • definitive quantitation of at least 10% samples; recommended for inorganics; determine precision, accuracy, and confidence limits on at least 1% samples. • both definitive identification and quantitation, as stated above.	Analyte specific. Definitive identification on 100% of the critical samples for organic analysis. Definitive quantitation and determination of confidence limits (precision and accuracy) on 100% of the critical samples.
Potential Applications	Physical/chemical properties of samples, extent/degree of contamination, pollutant plume definition in ground water, monitor well placement, waste compatibility, preliminary health and safety assessment, hazard characterization, preliminary identification of pollutants.	Physical/chemical properties of samples, extent/degree of contamination, verification of pollutant plume definition in ground water, verification of health and safety assessment, verification of pollutant identification, verification of cleanup.	Make a decision, based on the action level, with regard to treatment, disposal, site remediation and/or removal of pollutants, health risk or environmental impact, responsible party identification, enforcement actions, cleanup verification.

QC and Data Validation Requirements		
Instrument calibration or a performance check of a test method. Detection limit should be determined, unless inappropriate.	Sample holding times. Method blanks. Rinsate blanks, if dedicated sampling tools are not used. One rinsate blank per parameter per 20 samples. One trip blank (Two 40ml vials filled with distilled/deionized water) per cooler of volatile organic samples. Performance evaluation sample (optional) Definitive identification on 10% samples. Definitive quantitation by reanalyzing 10% samples; determination of precision, accuracy, and confidence limits by analyzing 10% or 2 pairs of matrix spike duplicates (whichever is greater). Initial and continuing calibration data. Detection limit should be determined, if appropriate.	Sample holding times. Initial and continuing calibration data. Definitive identification via a Gas Chromatograph column or Mass Spectrograph on 100% samples for organics. Definitive quantitation for 100% of the critical samples; determination of precision, accuracy, and confidence limits by analyzing 20% or 4 pairs of matrix spike duplicates (whichever is greater). Method blanks, rinsate blanks, trip blanks, as in QA2. Performance evaluation samples. Detection limit should be determined. Review of 10% samples for all the listed elements; review holding times, blank contamination, precision, accuracy, error determination, detection limits, and confirmed identification for the remaining 90%. Review of all elements for all samples in each analyte category for every 10th data package.

Using the QA/QC Objective. As the QA/QC Objective reflects the importance of and the intended use of the data, it also can be used to assist the OSC in the determination of the sampling methodology and the total number of field and control samples. In addition, it is also useful in the determination of the analytical methodology and the data validation criteria.

The OSWER Directive 9360.4-01 provides some guidelines to enable the OSC to perform these tasks; however, the OSC has to use substantial judgement in performing them. In particular, the determination of the number of samples and the production of the chain of custody forms can be reasonably well defined by associating simple rules of thumb with the selected QA/QC Objective. On the other hand, IQAP, described in the sections below, is designed to cater to the difficult need of specifying the analytical methodology.

Inputs and Outputs

This section describes the inputs and the outputs of IQAP.

Inputs. For the design of IQAP system, it is assumed that the inputs are obtained primarily in the selection of the QA/QC Objective. These inputs include:

o   the objective of the sampling event, e.g., collect groundwater samples

o   the intended use of the generated data, e.g., check if utility wells need to be closed

o   a description of the waste materials that are likely to be handled at the site, including the matrix, likely concentration ranges, and volumes

o   the QA/QC Objective, and related QA/QC requirements (Table 1) including the number of samples, which will be obtained in the future with assistance from ERT's expert system being developed to select the QA/QC Objective

o   the equipment that has to be employed for testing for the specified QA/QC Objective, and a specified level of sensitivity [OSWER Directive 9360.4-01 (3) notes that the concentration level, spedific or generic, is needed in order to make an evaluation and for determination of the analytical method to be used. It is often the action level 1.2., the concentration at which removal actions have to be undertaken.].

Outputs. The primary output of IQAP are the correlation of detailed QA/QC requirements (criteria). By specifying the QA/QC criteria, IQAP will enable the OSC to tailor analytical methods and margins of error for his site-specified needs. Transmission of these data, along with the samples' information, and the validation of the data, are also activities that can be aided by computer based automated systems.

The IQAP Knowledge Processor

The selection of the QA/QC Objective requires expertise in assessing data quality, in predicting the impact of analytical results, and in relating the sampling process with the intended data use objective. On the other hand, the use of the QA/QC Objective requires expertise in chemistry so that it can be translated into a sound analytical methodology. IQAP is designed to embed this expertise into a computer-based expert system. However, to do so, one requires a knowledge representation strategy, i.e., the appropriate formalism that will provide the programming modules and templates needed to describe and replicate the experts' reasoning and problem solving methods inside the computer. The sections below describe the functional requirements for the knowledge representation strategy and processing characteristics of IQAP.

Knowledge Representation for IQAP. Three knowledge representation schemes, i.e., relational, rule-based and frame-based, and constraint-based, were examined for developing IQAP's knowledge base. For a general description of some knowledge representation schemes (4).

Relational Representation Scheme. The Quality Control (QC) criteria for environmental samples can often be stated (represented) in a set of tables and matrices. For example, the CLP diskette deliverables' format is indicative of an effort to provide such a generic representation. In fact, a well designed relational structure can provide a reasonably useful generic representation for the QC criteria. However, the structure of the relational form is not easily modifiable, and cannot capture easily the numerous ways in which chemists like to specify and use QC criteria. For example, within the relational framework, it takes significant design and programming effort to modify the range of acceptable relationships between the responses for the various masses in the GC/MS tuning and MS calibration criteria during a volatile organics analysis. To summarize, the relational form can be useful in encoding and maintaining standard QC criteria that do not change much, such as CLP criteria.

Rule-based or Frame-based Representation Scheme. A rule-based or frame-based representation of the QC criteria may make it easier to specify, use, and modify the QC criteria. In a rule-based representation, the QC criteria will be specified in an "IF <condition> THEN <action>" form. For example, a QC criterion may be stated as "IF <extraction performed in less than 7 days after sampling> THEN <holding time is OK>." In contrast, in a frame-based system, a data element called "sample" will have property slots called <extraction time> and <holding time>. The <extraction time> property slot will be programmed to recognize that when a user assigns a value, say "6" days, to the <extraction time> property slot, the <holding time> property slot will be set to "OK." A rule-based or frame-based system can be utilized by the user to make small modifications to an existing set of criteria, such as the CLP criteria. Such a system can be designed to interact with a relational database of the standard QC criteria.

The advantage of these systems over the procedural, relational approach is that the experts often do not have all the rules of thumb for each and every situation. A rule-based or frame-based approach will enable the developer to build a system to replicate the state of knowledge as it exists, with an inherent ease of enhancement by addition or modification of the rules. However, the number of rules needed to enable the system to perform these tasks in a generic manner for any number of site conditions will be inordinately large. Further, once such a system is designed, it is not always easy to resolve subsequent conflicts among rules that are added to the system.

To summarize, a rule-based or frame-based system could provide a reasonably flexible tool to the Superfund Program, which may at times need to specify and utilize small modifications of a standard, such as the CLP QC criteria. Such a system could select a CLP method and its associated QC criteria based on the inputs (Section 3.1), modify the QC limits for certain analytes and/or samples, and specify modifications in the rules of thumb used for validation. However, the potential for an inordinately large number of rules, this apparently simple solution has the practical drawback of embedding conflicts among the rules which will be difficult to debug and modify. This is contrary to the way in which experts think and consequently, this scheme is deemed to be unsuitable.

Constraint-based Representation Scheme. The primary feature of the chemists' ability to specify QC criteria based on the QA/QC Objective and other inputs is that the QC criteria are specified as a set of general constraints and limits which the data must adhere to. For example, for a PCB site with a QA/QC Objective level of QA2, the experts could specify constraints for the data such as "the matrix spike duplicate recoveries should be between 80% to 120%", and that "the PCB detection level must be at least 1.5 ppm." Therefore, a constraint-based representation presents a natural mechanism to emulate the thought process of the chemists in developing QC criteria.

A constraint-based system may offer some interesting mechanisms to enable the chemist to update and modify the system. For example, a high level (more generic) constraint may be stated to be "the QC parameters should be related to some action level and the inputs of the system." Subsequently, the knowledge base can associate various constraints such as "<QC parameter> <operator> <action level> <QA/QC Objective and analytical method>" based on the experts' experience, e.g. "<PCB concentration> <less than> <1.5 ppm> <QA2 and GC>." No matter how many such constraints are then developed or changed based on the experts' rules of thumb, the generic statement that "the appropriate constraints for the site should be satisfied" can be applied. For example, criterion for other sites may be modeled by adding a low level (more specific) constraint to be "<PCB concentration> <less than> <1.0 ppm> <QA3 and GC>" or by changing the high level (more generic) constraint to be "<QC parameter> <less than or equal to> <action level> <QA/QC Objective and Equipment type>." Upon processing such constraints, their intersection can be used and specified in a relational form.

This allows the expert to specify his constraint at the level of detail at which they are applicable in the form in which he likes to think of them. The ease with which the system can incorporate the form of the experts' thought process and its ability to upgrade its state of knowledge without a complete overhaul is a good indication that this knowledge representation scheme may deserve closer attention.

Processing Characteristics of IQAP.   To follow the experts' methodology, it appears that the processing within IQAP will be performed in two steps. In the first step, the inputs should be accepted and a constraint-based system will develop a set of constraints and limits that apply to the data. The knowledge embedded in this first system will convert the constraints into an analytical method and the associated set of QC criteria. It is expected that because QA/QC Objectives incorporate considerations regarding the usability of the data, the constraints will be flexible enough to process the inputs into a reasonably structured set of data tables. Therefore, in the second step, the method and its associated QC criteria will be represented in a relational database. The level of detail in the specifications of the QC criteria and the relationships between the data elements will be more specific as the process proceeds from the first step to the second step.

Conclusions

A hybrid knowledge representation strategy appears promising for the implementation of chemists' knowledge for developing site-specific QA/QC plans. The primary knowledge representation strategy is a constraint-based scheme which interprets the inputs to develop a hierarchy of QC constraints that may apply to the site. The constraints are processed by translating them into the static structure/format of a relational database of QC criteria.

References

1. Guidelines and Specifications for Preparing Quality Assurance Program Plan, U.S. Environmental Protection Agency, 1980, QAMS-004/80.

2. Interim Guidelines and Specifications for Preparing Quality Assurance Project Plans, U.S. Environmental Protection Agency, 1980, QAMS-005/80.

3. Removal Program Quality Assurance/Quality Control Interim Guidance, U.S. Environmental Protection Agency, 1989, OSWER Directive 9360.4-01.

4. Rich, E. Artificial Intelligence, McGraw-Hill, Inc., New York, NY. 1983.

RECEIVED April 13, 1990

Chapter 7

An Expert System for Prediction of Aquatic Toxicity of Contaminants

James P. Hickey, Andrew J. Aldridge, Dora R. May Passino, and Anthony M. Frank

U.S. Fish and Wildlife Service, National Fisheries Research Center—Great Lakes, 1451 Green Road, Ann Arbor, MI 48105

The National Fisheries Research Center-Great Lakes has developed an interactive computer program in muLISP that runs on an IBM-compatible microcomputer and uses a linear solvation energy relationship (LSER) to predict acute toxicity to four representative aquatic species from the detailed structure of an organic molecule. Using the SMILES formalism for a chemical structure, the expert system identifies all structural components and uses a knowledge base of rules based on an LSER to generate four structure-related parameter values. A separate module then relates these values to toxicity. The system is designed for rapid screening of potential chemical hazards before laboratory or field investigations are conducted and can be operated by users with little toxicological background. This is the first expert system based on LSER, relying on the first comprehensive compilation of rules and values for the estimation of LSER parameters.

The National Fisheries Research Center-Great Lakes, U.S. Fish and Wildlife Service, has tentatively identified more than 500 contaminants in the tissues of walleyes (Stizostedion vitreum vitreum) and lake trout (Salvelinus namaycush) from the Great Lakes basin (1), and 362 substances have been verified in the Great Lakes system (2). A systematic assessment of the biological hazards of these compounds is underway (3-5). It is, however, physically and economically impossible to run bioassays on every compound, especially since hundreds of new compounds are being introduced into the environment each year.

By using mathematical models based on quantitative structure-activity relationships (QSAR), one can quickly determine the compounds that merit further investigation (5-9). Use of these predictive models serve as a rapid, inexpensive screening technique, especially for compounds not commercially available. Such models, however, are complex and much of the knowledge required to apply them is qualitative or not well defined. One approach to making these models usable as an assessment tool for toxicologists, in lieu of a bioassay, or managers who lack technical expertise in toxicology and chemistry is to create a computer program capable of using qualitative information for decision-making (a so-called expert system). Expert systems are special-purpose programs that imitate the performance of human experts in solving problems on specialized, often highly technical subjects (10); they do so using the heuristic knowledge of the human expert, with the expert's qualitative reasoning to solve the problem (11). Expert systems have potential for recognizing and managing environmental problems. Prototype expert systems being developed include site assessment systems used to identify and quantify hazards at "superfund" sites. Also included are predictive systems such as the Hazardous Waste and Management Expert System designed to provide advice about reducing health and environmental risks at hazardous waste sites (12).

We developed the first expert system that incorporates a working set of rules for a type of QSAR referred to as a linear solvation energy relationship or LSER (13-17) to predict LSER variable values from SMILES string formalism. The program also uses these LSER results and information about toxicity to predict acute toxicity to four representative organisms: the fathead minnow (Pimephales promelas), the crustaceans Daphnia magna and Daphnia pulex, and Photobacterium phosphoreum, the luminescent agent in the Microtox test.

System Overview
The expert system is designed to predict LSER variable values from a chemical structure formalism, and an addition to the software uses these values to predict acute toxicity (Figure 1). The program is written in the muLISP computer language and runs on an IBM-compatible personal computer. It can be used by anyone who has even a limited background in toxicology or chemistry. It adheres to a basic doctrine of expert system methodology: the separation of the knowledge base from the methods of processing the knowledge makes the system relatively easy to modify and debug (10). Expert systems are designed to be used by non-technical personnel (18); in our example the ultimate user could be a natural resource manager. To operate our system, the user enters a representation (structure or suitable identification number) of a chemical compound into the computer. No other interaction is required.

The program can be divided into three main sections: the first section determines the structural elements of the chemical compound; the second estimates the values for the LSER model variables; and the third predicts the toxicity of the compound. All three sections are controlled by the main reasoning section, called the inference engine.

To identify the compound the inference engine queries the user for either a CAS (Chemical Abstracts Service) identification number or a SMILES representation of the compound structure (defined later). Next, the string is decomposed by a fragmentation process and two rule bases are consulted. The inference engine constructs a secondary internal knowledge base using these fragments by forward chaining of

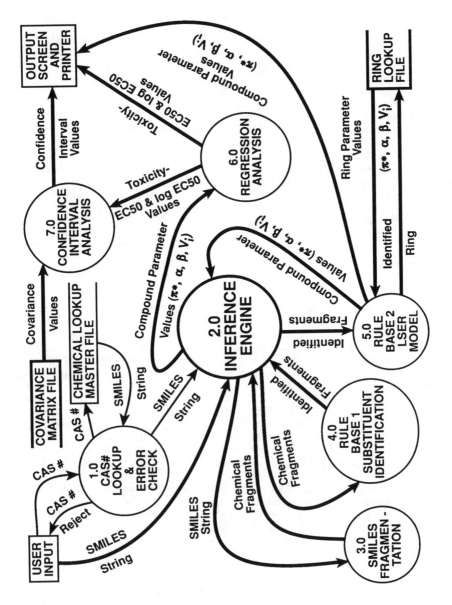

Figure 1. Physical Data Flow Diagram.

the rule bases and stores the information in a dynamic, global, intrinsic data type characterized by LISP programmers as a "property list". Values for the four LSER variables ($V_i/100$, π^*, β, α, vide infra) are assigned to the individual fragments of the compound by the second rule base and accumulated. The inference engine then invokes the regression and confidence interval calculations, utilizing the LSER variables as inputs. Finally, the estimated LSER variables and the predicted toxicities with their confidence intervals are displayed to the user.

SMILES Strings
The expert system determines the structure of an organic molecule from a standard chemical notation known as a SMILES string. The $\underline{S}$implified $\underline{M}$olecular $\underline{I}$nput $\underline{L}$ine $\underline{E}$ntry $\underline{S}$ystem was developed by the U.S. Environmental Protection Agency (19, 20) at the Environmental Research Laboratory, Duluth, Minnesota, for the QSAR Research Program (20-23) and was based on work with the Medicinal Chemistry Project at Pomona College, Claremont, California (24). A SMILES string is a linearization of the three-dimensional representation of an organic compound. The notation has four basic syntax rules that allow the user to represent a molecule in a form that can be easily used by the muLISP language (19-21). A synopsis of the rules will be given here for the prospective user; in the future, the use of the SMILES formalism will not be so necessary.

The first rule designates a set of basic symbols (Table I). All molecules are represented as hydrogen-suppressed, and single bonds are assumed by default. For example, CO assumes a single bond between the carbon and oxygen, and C=O indicates a double bond. The bond between two lowercase symbols is aromatic.

Table I. Basic Symbols Used in the Formulation of SMILES Strings

Symbol	Designation	Symbol	Designation
C,c	Normal and aromatic carbon	BR,Br	Bromine
N,n	Normal and aromatic nitrogen	CL,Cl	Chlorine
O,o	Normal and aromatic oxygen	I,F	Iodine and fluorine
S,s	Normal and aromatic sulfur	=,#	Double, triple bonds
P,p	Normal and aromatic phosphorus	*	Aromatic bond

The second rule defines simple chains in molecules. A simple chain of atoms is represented by atomic symbols interspersed with their respective bond symbols. For example, CC represents ethane; C=C ethene; and CCCCCO, n-pentanol.

The third rule defines simple branches in molecules. A branch from the main chain is enclosed in parentheses. The string in parentheses is placed directly after the symbol for the atom to which the branch is connected. If it is connected by a multiple bond, the bond symbol immediately follows the left parenthesis. More than one

branch is indicated by using more than one set of parentheses: ( )
( ) and (( )) are simple forms that may be used. For example, IC(I)O
represents   diiodomethanol,   and   CCCCC(C(C)C)CC=C   represents   3-
isopropyloctene.
     The fourth rule defines ring structures. A ring is closed by using
a pair of ring closure numbers. In ClCCCCCl, a single bond connects
the "l" after the first carbon with the other carbon followed by a
"l".   For multiple rings, we use different ring numbers (e.g.,
1,2,3,4), and pairs of carbons with like numbers are connected to
close the rings. For example, clcc2ccccc2ccl represents napthalene,
where cl is joined with cl and c2 with c2 to form the two fused rings.

Inference mechanics
The problem-solving system is built around rules that consist of an
antecedent "if" part and a conclusion "then" part. These rules are
processed by concentrating on the rules' antecedents, a process
referred to as forward chaining (11).
     When all of the antecedents in a rule test true, the rule is said
to be triggered. If an action is performed (i.e., a conclusion is
added to the secondary knowledge base), the rule is said to be fired.
Several rules may be triggered at once, requiring conflict resolution
strategies to determine which of them should be fired (25).
     The conflict resolution strategies we use are termed context
limiting and rule ordering. In context-limiting strategy, rules are
grouped in such a way that few rules are active at the same time.
The resulting groups are disjoint subsets of the set of all rules.
Because the inference engine activates and deactivates rule groups,
it needs only to handle conflicts within a group. Within each subset,
rule ordering is used to resolve conflicts. Rule ordering requires
that the rules be organized in a single priority list in which the
first triggering rule in the list has the highest priority and the
others are ignored (25).
     The inference engine isolates the basic skeletal structures, (rings
or chains) and all other functional groups composing the compound, and
passes these to the first knowledge base that identifies them. Rings
are the most difficult structures to isolate because one compound may
contain many rings with some attached to each other. The SMILES
string is therefore put in a tree structure, and the shortest path to
the numbers in the chain are identified by using a branch and bound
search (25).
     For example, Figure 2a shows a tree for decahydro-2,3-
dimethylnaphthalene   for   which   the   SMILES   string   is
(C(C(CCCl)CC(C2C)C)(Cl)C2). The inference engine first finds the "l"
ring and then proceeds to the "2" ring. Once the shortest path to one
of the numbers of the numbered ring is found, the system balances the
tree and runs the search again. The first number found becomes the
root (Figure 2b). This shortest path then becomes the actual ring
substituent, which is stored for later identification by the first
rule base in the secondary knowledge base.
     The inference engine then begins another search for another ring,
repeating the search sequence. All non-ring fragments are also
isolated and stored in the knowledge base of the secondary property

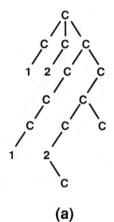

(a)

Figure 2a. SMILES string tree representation for decahydro-2,3-dimethylnaphthalene, C(C(CCC1)CC(C2C))(C1)C2.

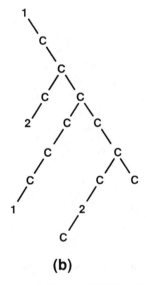

(b)

Figure 2b. SMILES string tree representation for SMILES string tree from a with C1 as root.

list. Only after all fragments of the chemical compound have been
isolated and stored in a property list does the program progress to
the first knowledge base for fragment identification. The software is
coded to recognize a certain large but finite number of fragments (or
functional groups).   If a particular functional group is not
recognized, the system will identify parts of it and proceed. If a
functional group is recognized and no values are available, the system
skips them, effectively assigning default values for the atoms making
up the unit. Once the substituents have been identified and their
names stored in the property list, the program progresses to the
second knowledge base used to generate the four LSER variable values.

Linear Solvation Energy Relationship
In the LSER model (13-17) the toxicity of a contaminant is related to
its structure (26).   The generalized LSER equation contains three
simple and conceptually explicit types of terms:

toxicity = cavity term + dipolar term + hydrogen bonding terms

In this system each fragment of the contaminant molecule
contributes both to the energy required to order solvent molecules
(water or biosystem medium) around the molecule and to the energies
gained or lost through formation of electrostatic and hydrogen bonds
between the contaminant and the medium. The general form of the
equation used in our expert system is

$$\log_{10}(\text{toxicity}) = \underline{m}V_i/100 + \underline{s}\pi^* + \underline{b}\beta + \underline{a}\alpha$$

where m, s, b, and a are constants. Numerical values of four LSER
variables are generated for each fragment: $mV_i/100$ is an endoergic
energy term that measures the free energy required to separate solvent
molecules and provide a suitably sized cavity for the contaminant
molecule, and $V_i/100$ is the intrinsic (van der Waals) molecular volume
scaled by a factor of 100, to give magnitudes comparable to the other
three variables. The dipolarity/polarizability term $s\pi^*$, measures the
generally exoergic effects of solute-solvent, dipole-dipole, and
dipole-induced dipole interactions, and π^* is a measure of the
molecule's ability to stabilize a neighboring charge or dipole by
nature of non-specific dielectric interactions. The hydrogen bonding
terms $b\beta_m$ and $a\alpha_m$ measure the exoergic effects of hydrogen bonding,
involving the solvent as hydrogen bond donor acid and the solute as
hydrogen bond acceptor base β_m, and the solute as hydrogen bond donor
acid and the solvent as hydrogen bond acceptor base α_m.
    Essential to the program are the complete sets of variable values
for each fundamental structure and fragment we have encountered or
that we anticipate will exist in an environmental sample. Some of
these values have been formulated by the few published rules (27, 28),
but most were computed primarily by extrapolation from other values
taken from the literature (6, 7, 13-17, 27-41) and codified into
rules.   A manuscript listing the comprehensive set of rules is in
preparation. Several conventions of the volume term are available

(39-42). The present system uses the convention V_i and makes use of contributions for all fragments and fundamental structures based on extrapolations from previously reported values (6, 7, 13-17, 27-41). The same process was used to devise value contributions of the π^*, β, and α variables. The present complete set of variable estimation rules allows prediction of the LSER variables for almost any organic compound. With regards to the accuracy of these estimated values, predictions for $V_i/100$ are generally ±.02 of literature values, as volumes are strictly additive. For α and β, the limited experimental data available from M. H. Abraham et al. (43, 44) show that the predicted values generally agree within ±.03 of the experimentally determined data. For π^*, there are no experimental data available, but predicted values agree ±.03 with published values (6, 7, 13-17, 27-41).

Bioassay Data Sets and Multiple Linear Regression Equations

The expert system predicts the acute toxicity of a chemical to four representative aquatic organisms and reports toxicity as either EC50-
-is the effective concentration at which either 50% of the animals (Daphnia pulex or D. magna) were immobilized or 50% of the luminescence (the Microtox test) was diminished--or LC50, the lethal concentration for 50% of the fish (fathead minnows) in the study.

The regression equations were derived by using one set of toxicity data collected under controlled conditions for each species. The data for Daphnia pulex (7) were obtained at the National Fisheries Research Center-Great Lakes, under the test conditions of a 48-hour exposure at 20°C in reconstituted hard water. The EC50 values were determined by probit analysis. The data sets for the Microtox test, Photobacterium phosphoreum (36), Daphnia magna (7), and fathead minnows, Pimephales promelas (45-48) were taken from the literature. All data sets were examined closely for continuity and applicability of test conditions, close adherence to rigid quality assurance and quality control schemes (49-52), and representative of a wide variety of chemical classes and structural subunits. Each LSER model is best developed by using a data set containing the widest selection of chemical classes and structures, which are generally representative of environmental samples. The regression equations used in the present system (Table II) were previously developed and discussed (7, 9, 36).

Results and Discussion

Our expert system is the only predictive software available based on the LSER model. This system also represents the first total codification of the rules and fragment contribution values for synthesis of the four parameter values. Previous publications have either listed partial guidelines for specific classes (27, 28) or have alluded to them (7, 13-17, 27-41). Table III demonstrates the predictive ability of the software for both the LSER parameters and contaminant toxicity. The LSER parameter values are composed from a sum of the contributions from each of the fragments, and the predicted values are generally close to the values composed by hand. For some compounds (such as acenaphthene), the values for π^*, β, and α may not be accurately represented by a simple sum of fragment variable

Table II. Expert System Multiple Linear Regression Equations

General Equation[a]

$$\text{Log}_{10}\ (\text{toxicity}) = (\text{intercept}) - mV_i/100 - s\pi* + b\beta - a\alpha$$

---

1) The Microtox Test (<u>Photobacterium phosphoreum</u>)[b], μM/L

$$\log(\text{EC50}) = 7.49 - 7.39\ V_i/100 - 1.38\ \pi^{\cdot} + 3.70\ \beta - 1.66\ \alpha$$
$$N = 40,\quad R^2 = 0.966,\quad sd = 0.319$$

2) <u>Daphnia pulex</u>[c], μM/L

$$\log(\text{EC50}) = 4.09 - 4.33\ V_i/100 - 0.05\ \pi^{\cdot} - 0.13\ \beta - 0.22\ \alpha$$
$$N = 38,\quad R^2 = 0.868,\quad sd = 0.418$$

3) <u>Daphnia magna</u>[c], mM/L

$$\log(\text{EC50}) = 4.18 - 4.73\ V_i/100 - 1.67\ \pi^{\cdot} + 1.48\ \beta - 0.93\ \alpha$$
$$N = 53,\quad R^2 = 0.948,\quad sd = 0.221$$

4) Fathead minnow (<u>Pimephales promelas</u>)[d], M/L

$$\log(\text{LC50}) = -0.34 - 5.26\ V_i/100 - 0.80\ \pi^{\cdot} + 3.98\ \beta - 0.80\ \alpha$$
$$N = 76,\quad R^2 = 0.970\quad sd = 0.218$$

---

a. see text for explanation of symbols.
b. (<u>36</u>)
c. (<u>7</u>)
d. (<u>9</u>)

Table III. Predicted (P) vs Experimental (E) LSER Parameter Values
and Acute Toxicities for the Microtox Test (MT), <u>Daphnia</u> <u>pulex</u> (DP),
<u>Daphnia</u> <u>magna</u> (DM), and the Fathead Minnow (FM)

Compound	P/E[c]	V_i	π'	β	α	MT EC50 (μM)	DP EC50 (μM)	DM EC50 (mM)	FM LC50 (M)
		100							
Well-behaved compounds[c]									
n-Hexanol	P	0.633	0.40	0.42	0.35	3.23	2.12	0.81	-2.63
	E	0.690	0.40	0.45	0.35	2.71		0.32	-3.02
n-Heptanol	P	0.731	0.40	0.42	0.35	2.51	1.69	0.35	-3.15
	E	0.789	0.40	0.45	0.33	1.93		-0.09	-2.53
2-Butanone	P	0.478	0.65	0.48	0.00	4.84	2.85	1.54	-1.51
	E	0.477	0.67	0.48	0.00	4.85		2.09	-1.35
4-Methyl-2-pentan-one	P	0.674	0.65	0.48	0.00	3.39	2.00	0.61	-2.54
	E	0.663	0.63	0.48	0.00	2.90		1.17	-2.30
1,2-Dichloroethane	P	0.376	0.70	0.20	0.00	4.49	3.26	1.52	-2.10
	E	0.442	0.81	0.10	0.00	4.05		1.13	-2.92
Iodocyclohexane	P	0.779	0.32	0.05	0.00		1.50		
	E	0.779	0.32	0.05	0.00		1.53		
Cyclohexane	P	0.598	0.00	0.00	0.00	3.07	2.30	1.35	-3.48
	E	0.598	0.00	0.00	0.00			0.61	-4.27
OMTA[d]	P	1.544	0.06	0.00	0.00		-1.80		
	E	1.444	0.04	0.00	0.00		-1.74		
Benzene	P	0.491	0.59	0.10	0.00	3.42	2.75	1.01	-3.01
	E	0.491	0.59	0.10	0.00	3.31		1.16	-3.40
o-Xylene	P	0.687	0.59	0.12	0.00	2.04	1.90	0.12	-3.96
	E	0.671	0.51	0.12	0.00	1.94		0.15	-3.82
Chlorobenzene	P	0.581	0.71	0.07	0.00	2.48	2.35	0.34	-3.69
	E	0.581	0.71	0.07	0.00	2.12		0.44	-3.77
Naphthalene	P	0.753	0.70	0.15	0.00	1.51	1.61	-0.33	-4.28
	E	0.753	0.70	0.15	0.00		1.58		-4.32
9H-Fluorene	P	0.958	0.66	0.21	0.00		0.68		
	E	0.960	0.66	0.20	0.00		0.11		

Table III. *Continued*

Compound	P/E[c]	LSER values[a]				log(Acute Toxicity)[b]			
		V_i	π	β	α	MT	DP	DM	FM
		100				EC50 (μM)	EC50 (μM)	EC50 (mM)	LC50 (M)
Troublesome compounds									
Acenaph-	P	0.855	1.83	0.30	0.00	-0.24	1.13	-2.49	-5.10
thene	E	0.896	0.62	0.17	0.00				-4.95
Camphor	P	1.594	0.68	0.48	0.00	-3.11	-1.79	-3.58	-7.16
	E	1.106	0.68	0.59	0.00				*-3.92
Diethyl-	P	1.177	0.93	0.82	0.00	0.10	-0.16	-1.91	-4.55
phthalate	E	1.153	0.90	0.70	0.00			*-0.52	-3.87
2,4-Pen-	P	0.662	1.30	0.96	0.00	4.36	2.08	0.29	-1.13
tanedione	E	0.595	0.90	0.90	0.00			0.00	*-2.98
Phenol	P	0.536	0.72	0.33	0.61	2.74	2.46	0.36	-2.94
	E	0.536	0.72	0.33	0.60	2.63		-0.48	-3.94
2-Methyl-	P	0.634	0.72	0.34	0.61	2.06	2.03	-0.09	-3.42
phenol	E	0.634	0.70	0.33	0.57	2.28		-0.75	-3.77
Aniline	P	0.562	0.71	0.50	0.23	3.83	2.44	0.86	-2.10
	E	0.562	0.73	0.50	0.26			*-2.27	-2.84
4-Chloro-	P	0.652	0.83	0.47	0.23	2.88	2.04	0.19	-2.79
aniline	E	0.652	0.73	0.40	0.31			*-1.59	-3.62
4-Fluoro-	P	0.591	0.74	0.45	0.28	3.46	2.31	0.62	-2.40
aniline	E	0.591	0.73	0.47	0.23*				*-3.82
4-Nitro-	P	0.702	1.13	0.70	0.23			-0.21	-2.40
aniline	E	0.702	1.25	0.48	0.42			-0.76	-3.04
Butylamine	P	0.472	0.25	0.69	0.00	6.21	2.93	2.55	-0.33
	E	0.535	0.32	0.69	0.00			*0.02	*-2.44
Trietha-	P	0.803	1.35	1.95	1.05	5.16	1.39	0.03	1.10
nolamine	E	0.840	1.35	2.00	0.85			0.97	*-1.10
Pyridine	P	0.470	0.87	0.44	0.00	4.44	2.87	1.15	-1.80
	E	0.470	0.87	0.44	0.00	4.51		0.48	-2.93
Nicotine	P	1.041	1.01	1.14	0.00		0.18		
	E	0.975	1.01	1.17	0.00		*1.34		

Continued on next page

Table III. *Continued*

---

a. See text for definitions of LSER symbols.
b. * Designates actual toxicity values greater than 2 or "sigma" from the predicted toxicity.
c. P = predicted values when expert system is used.
    E = hand-calculated LSER parameter values or experimentally determined toxicity for a species.
    For hand-calculated LSER variable values, the contributions from all fragments were summed. Some of the values for π^*, β, and α were adjusted to reflect either some predominant contribution or a vector sum, which can account for most of the discrepancies between predicted and experimental values.
    Blank toxicity entries (P or E) indicate no data were available.
    "Well Behaved Compounds" have LSER system-predicted values that agree ±.01 with hand-calculated values and predicted toxicities within ±1 log unit of actual value.
    "Troublesome Compounds" can have LSER values >±.03 with hand-calculated values but generally have actual toxicities ±1 to 3 log units different from the predicted values.
d. OTMA: octahydro-1,4,9,9-tetramethyl-1H-3a,7-methanoazulene

---

contributions (which is the present situation); a vector sum or a sum with a fragment hierarchy of importance (use or not use, and to what degree) may give a more accurate value. This approach is done occasionally to compute hand-calculated values found in Table III and accounts for the discrepancies occasionally seen between predicted and experimental values for LSER variables.

Even when the expert system's estimates of LSER parameter values match our hand calculations, predicted toxicities can differ from the observed toxicity by one to three orders of magnitude (see "Troublesome Compounds" in Table III). Our expert system was developed to predict toxicities according to one mode of action-- nonpolar, non-reactive narcosis-- for neutral organic molecules with no special physical property considerations. Toxicities for such compounds have been predicted by our system within one order of magnitude of the observed value and generally within ± a factor of 5. Compounds that react with the biosystem (e.g., aldehydes and amines) generally are 10 to 1000 times more toxic than predicted by the present system; and compounds that ionize at test pH conditions (organic acids), that are highly volatile (e.g., camphor), that have low water solubility, or that do not diffuse across cell membranes (very long chain alcohols) will have an observed toxicity of only 1/10 to 1/100 of that predicted here. Toxicity would also be difficult to estimate for organic esters (like phthalates) and amides because hydrolysis under the test conditions would produce at least two different, possibly toxic, molecules (19).

Expert System Capability, Utility, and Limitations
The expert system is capable of evaluating compounds with either ring or chain skeletons, double and triple bonds, and all common functional groups (e.g., acids, alcohols, amines, esters, and halides). The ring structures include cyclohexane and benzene derivatives; multiple, and condensed ring structures such as polychlorinated biphenyls (PCBs), naphthalenes, and higher polynuclear aromatic hydrocarbons and their heterocyclic analogs such as nicotine and pyrrole; and the corresponding saturated ring systems.

The LSER models provide consistently better correlations with toxicity than do other widely used QSAR models that depend simply on the partitioning of contaminants into octanol and water, K_{ow} (28). The LSER models have thus far been applied to only a few data sets for aquatic organisms (7, 9, 36-38) although they could be used to predict toxicity to a wide variety of aquatic organisms, as well as to model specific mechanisms of toxicity (37, 38).

We are currently using the expert system to help estimate toxicity of chemicals before we begin bioassays, to shorten the time spent on range-finding tests. The system also may be used as part of a hazard assessment scheme, to evaluate the toxicity of compounds detected in environmental samples by gas chromatography/mass spectrometry (GC/MS). The system could act as a screening tool in an initial evaluation of contaminants detected at a site of concern. Some limitations of the system in chemical recognition and in the estimation of values of LSER variables must be improved. As in any viable, evolving expert system, there are still software problems and limitations being studied. For example, certain SMILES designations cannot be analyzed correctly, and certain obscure chemical fragments (class types) are identified incorrectly. Also, there is no provision to differentiate between possible geometric (cis vs trans) or optical isomers of compounds. The different forms have different observed toxicities, but no applicable weighting scheme or input designation has been developed for the present system.

Future
The software now uses structurally intrinsic parameters for only one QSAR model (LSER) and the results are used to predict one property (acute toxicity) to four aquatic species by one mechanism (nonreactive, non-polar narcosis); however, we intend to continue to refine our equations as databases grow, incorporate other models, predict other properties, and include other organisms. We will attempt to differentiate between modes of toxic action and improve our estimates accordingly. For the widely divergent classes of chemicals and types of environmental behavior, no one model will best describe every situation and no one species is the optimal organism to monitor. As the software evolves, the expert system should choose the best model based on the contaminant, the species, and the property to be predicted (e.g., toxicity or bioaccumulation). In addition, we envision an interactive screen system for data entry that will bypass the SMILES notation and allow the user to describe the molecule by posing a series of questions about the compound's backbone and functional groups. The responses will translate directly into values of LSER variables.

A preliminary version of this system is now available. Our ultimate objective is to produce a user-friendly expert system for use in the evaluation of contaminants at specific sites.

Acknowledgments
We dedicate this work to the memory of the late Mortimer J. Kamlet whose inspiration and drive helped bring his LSER concept to a workable hypothesis that made this work possible. We thank Dr. Amjad Umar (University of Michigan) for concepts and guidance in artificial intelligence and Dr. Michael H. Abraham (University College London) for his consultation and encouragement in LSER theory and practice. Use of trade names does not constitute Government endorsement of commercial products.

Literature Cited
1. Hesselberg, R. J.; Seelye, J. G. Identification of organic compounds in Great Lakes fishes by gas chromatography/mass spectrometry: 1977. Administrative Report 82-1. Great Lakes Fishery Laboratory, Ann Arbor, Michigan. 1982.
2. Great Lakes Water Quality Board. 1987 Report on Great Lakes Water Quality. International Joint Commission, Windsor, Ontario. 1987. 236 pp.
3. Passino, D. R. M.; Smith, S. B. Environ. Toxicol. Chem., 1987, 6, 901-907.

4. Savino, J. F.; Tanabe, L. L. Bull. Environ. Contam. Toxicol., 1989, 42, 778-784.
5. Passino, D. R. M.; Smith, S. B. In QSAR in Environmental Toxicology; K. L. E. Kaiser, ed.; D. Reidel. Dordrecht, Holland, 1986, pp 261-270.
6. Passino, D. R. M. Proceedings of the Technology Transfer Conference, Ontario Ministry of the Environment, Toronto, Ontario, 1986, Part B, pp 1-26.
7. Passino, D. R. M.; Hickey, J. P.; Frank, A. M. Proceedings of QSAR 88: Third International Workshop on Quantitative Structure-Activity Relationships in Environmental Toxicology, 1988, pp 131-146.
8. Hickey, J. P.; Passino, D. R. M.; Frank, A. M. Preprint of Papers, 3rd Chemical Congress of North America and 195th ACS National Meeting, 1988, 28, 521-523.
9. Hickey, J. P.; Passino, D. R. M.; Kamlet, M. J. Program and Abstracts, 22nd Great Lakes Regional Meeting, American Chemical Society. 1989. Abstract No. 6.
10. Waterman, D. A. A Guide to Expert Systems, Addison-Wesley, Reading, MA, 1986.
11. Brundick, F.; Dumar, J.; Hanratty, T; Tanenbaum, P. In Second Conference on Artificial Intelligence Applications: The Engineering of Knowledge-Based Systems, 1985.
12. Hushon, J. M. Environ. Sci. Technol., 1987, 21, 838-841.
13. Taft, R. W.; Abboud, J.-L. M.; Kamlet, M. J.; Abraham, M. H. J. Solution Chem., 1985, 14:153-175.
14. Kamlet, M. J.; Doherty, R. M.; Abboud, J.-L. M.; Taft, R. W. Chemtech, 1986, 16, 566-576.
15. Kamlet, M. J.; Taft, R. W. Acta Chem. Scand. 1985, B39, 611-628.
16. Kamlet, M. J.; Abboud, J.-L. M.; Taft, R. W. Prog. Phys. Org. Chem., 1981, 13, 485-630.
17. Abraham, M. H.; Doherty, R. M.; Kamlet, M. J.; Taft, R. W. Chem. Britain, 1986, 22, 551-554.
18. Burns, N. A.; Ashford, T. J.; Iwaskiw, C. T.; Starbird, R. P.; Flagg, R. L. IBM Systems Journal, 1986, 25, 2.
19. Leo, A.; Weininger, D. CLOGP Version 3.2 User Reference Manual. Medicinal Chemistry Project, Pomona College. Claremont, CA, 1984.
20. SMILES: A line notation and computerized interpreter for chemical structures. Environmental Research Brief. U.S. Environmental Protection Agency, Environmental Research Laboratory, Duluth, MN, 1987.
21. Weininger, D. J. Chem. Information Computer Sci., 1988, 28, 31-36.
22. SMILES user Manual. Institute for Biological and Chemical Process Analysis (IPA), Montana State University, Bozeman, MT. 1987.
23. QSAR System User Manual, Montana State University, Bozeman, MT. 1987.
24. MedChem Software Manual. Medicinal Chemistry Project, Pomona College, Claremont, CA, 1985.
25. Winston, P. H. Artificial Intelligence, 2nd ed. Addison-Wesley, Reading, MA, 1984.

26. Casarett, L. J.; Doull, J. Toxicology, 3rd ed. C. D. Klaassen, M. O. Amdur, J. Doull, eds. MacMillan, New York, 1986.
27. Kamlet, M. J.; Doherty, R. M.; Abboud, J.-L. M.; Abraham, M. H.; Taft, R. W. J. Pharm. Sci., 1986, 75, 338-349.
28. Kamlet, M. J.; Doherty, R. M.; Carr, P. W.; Mackay, D.; Abraham, M. H.; Taft, R. W. Environ. Sci. Technol., 1988, 22, 503-509.
29. Kamlet, M. J.; Doherty, R. M.; Abraham, M. H.; Carr, P. W.; Doherty, R. F.; Taft, R. W. J. Phys. Chem., 1987, 91, 1996-2004.
30. Taft, R. W.; Abraham, M. H.; Famini, G. R.; Doherty, R. M.; Kamlet, M. J. J. Pharm. Sci., 1985, 74, 807-814.
31. Kamlet, M. J.; Doherty, R. M.; Abraham, D. J.; Taft, R. W.; Abraham, M. H. J. Pharm. Sci., 1986, 75, 350-355.
32. Kamlet, M. J.; Doherty, R. M.; Fiserova-Bergerova, V.; Carr, P. W.; Abraham, M. H.; Taft, R. W. J. Pharm. Sci., 1987, 76, 14-17.
33. Leahy, D. E.; Carr, P. W.; Pearlman, R. S.; Taft, R. W.; Kamlet, M. J. Chromatographia, 1986, 21, 473-478.
34. Sadek, P. C.; Carr, P. W.; Doherty, R. M.; Kamlet, M. J.; Taft, R. W.; Abraham, M. H. Anal. Chem., 1985, 57, 2971-2978.
35. Carr, P. W.; Doherty, R. M.; Kamlet, M. J.; Taft, R. W.; Melander, M.; Horvath, C. Anal. Chem., 1986, 58, 2674-2680.
36. Kamlet, M. J.; Doherty, R. M.; Veith, G. D.; Taft, R. W.; Abraham, M. H. Environ. Sci. Technol., 1986, 20, 690-695.
37. Kamlet, M. J.; Doherty, R. M.; Taft, R. W.; Abraham, M. H.; Veith, G.; Abraham, D. J. Environ. Sci. Technol., 1987, 21, 149-155.
38. Kamlet, M. J.; Doherty, R. M.; Abraham, D. J.; Taft, R. W. Quant. Struct. Activ. Relat., 1988, 7, 71-78.
39. Leahy, D. E. J. Pharm. Sci., 1986, 75, 629-636.
40. Pearlman, R. S. Partition Coefficient Determination and Estimation; W. J. Dunn, J. H. Block, and R. S. Pearlman, eds.; Pergamon Press, New York, 1986, pp. 3-20.
41. Bondi, A. J. Phys. Chem., 1964, 68, 441-449.
42. Abraham, M. H.; McGowan, J. C. Chromatographia, 1987, 23, 243-246.
43. Abraham, M. H.; Buist, G. J.; Grellier, P. L.; McGill, R. A.; Prior, D. V.; Oliver, S.; Turner, E.; Morris, J. J.; Taylor, P. J.; Nicolet, P.; Maria, P.-C.; Gal, J.-F.; Abboud, J.-L. M.; Doherty, R. M.; Kamlet, M. J.; Shuely, W. J.; Taft, R. W. J. Phys. Org. Chem., 1989, 2, 540-552.
44. Abraham, M. H.; Grellier, P. L.; Prior, D. V.; Duce, P. P.; Morris, J. J.; Taylor, P. J. J. Chem. Soc. Perkin Trans. II, 1989, 699-711.
45. Acute Toxicities of Organic Chemicals to Fathead Minnows (Pimephales promelas). Brook, L. T.; Geiger, D. L.; Call, D. J.; Northcott, C.E., eds. Center for Lake Superior Environmental Studies, University of Wisconsin, Superior, WI, 1984-1987, Vol. 1-4.
46. Veith, G. D.; Call, D. J.; Brooke, L. T. Can. J. Fish. Aquat. Sci., 1983, 40, 743-748.
47. Hall, L. H.; Kier, L. B.; Phipps, G. Environ. Toxicol. Chem., 1984, 3, 355-365.
48. Hall, L. H.; Kier, L. B. Environ. Toxicol. Chem., 1985, 5, 333-337.

49. Hamilton, M. A.; Russo, R. C.; Thurston, R. V. Environ. Sci. Technol., 1977, 7, 714-719. [Correction 1978, 12, 147].
50. American Public Health Association, American Water Works Association, and Water Pollution Control Federation. Standard methods for the examination of water and wastewater, 16th ed., New York, 1985.
51. Committee on Methods for Toxicity Tests with Aquatic Organisms. 1975. EPA-660/3-75-009. U.S. Environmental Protection Agency, Duluth, MN.
52. American Society for Testing and Materials. Annual Book of ASTM Standards, E729-80. American Society for Testing and Materials, Philadelphia, PA, 1980.

RECEIVED January 9, 1990

Chapter 8

A Citizen's Helper for Chemical Information

W. James Hadden, Jr.

Intelligent Advisors, Inc., 2400 Westover Road, Austin, TX 78703

The Emergency Planning and Community Right-to-Know Act of
1986 has fostered the development of a great deal of
intelligent software for planning against and responding
to emergencies, but little that serves its right-to-know
aspects. This paper describes a companion to the CAMEO
II system, which was developed at the behest of the U.S.
Environmental Protection Agency, that provides citizens
the opportunity to obtain information about hazardous
materials, their health effects, and facilities that use
or store them.

Environmental legislation took a new turn with the passage of SARA
Title III, the Emergency Planning and Community Right-to-Know Act of
1986. Instead of relying on regulatory agencies to set standards to
protect the environment, Title III assumes that citizens themselves
will help decide acceptable levels of risk from hazardous chemicals
in their communities. The provisions of the law focus on ensuring
that data about such chemicals are available; indeed, the complexity
of the reporting provisions makes computer use virtually a sine qua
non for effective use of the data.
    Existing computer programs that support Title III activities
have tended to focus on the needs of the reporting industry or, in a
few cases, upon the needs of users concerned about the emergency planning and response provisions of the law. This paper describes one
such system as well as a supplementary "Citizen's Helper" module
designed to implement the intent as well as the letter of the statute.
It is novel in its intention to provide intelligent assistance to a
lay, rather than a professional, audience.
    The first section of the paper describes the early version of the
emergency response software, a brief description of the law, and the
ways in which the existing software was modified to meet the needs of
the law. The second part describes the Citizen's Helper program,
focusing especially on features needed to ensure that citizens can
have access to and make use of the data provided under Title III.
Part 3 considers possible expansions of the Citizen's Helper as

0097–6156/90/0431–0108$06.00/0
© 1990 American Chemical Society

well as the general nature of intelligence and expertise in software
intended for lay rather than expert users.

Background

CAMEO (Computer Aided Management of Emergency Operations) was first
developed as an aid to people responding to emergencies involving
hazardous chemicals. Implemented in Business Filevision, the program
served three important purposes: 1) it provided users with information about procedures appropriate to each chemical, including protec-
tive gear and mitigation measures; 2) it allowed retrieval of this
information using any one of several synonyms for the chemical through
a separate module called "Codebreaker;" and 3) it calculated dispersion of chemicals in the air to provide responders with information
about the need for evacuation as well as mitigation actions. CAMEO
was developed by the Hazardous Materials Laboratory of the National
Oceanographic and Atmospheric Administration (NOAA) with funding from
EPA.

SARA Title III and CAMEO II. In October 1986, Congress passed the
Emergency Planning and Community Right-to-Know Act of 1986, commonly
known as SARA Title III because it is the third title of the Superfund
Amendments and Reauthorization Act. Title III is a complex law which
specifies several different reports, four different sets of chemicals,
three different sets of covered facilities, and several different
reporting dates. There are three levels of administration: 1) EPA,
which oversees the actions of and receives aggregated data reported
by 2) the State Emergency Preparedness Committees (SERCs), which
receive data concerning emissions of hazardous materials from
facilities and aggregated storage and use data from 3) Local Emergency
Planning Committees (LEPCs), whose jurisdictions were established by
the SERCs. Each LEPC is required under Title III to receive reports
from facilities that manufacture, store, or use chemicals on one of
the lists, to conduct emergency planning, and to oversee citizen
access to the data. (It should be noted that the LEPCs have been
given the responsibilities of a first-born son while being granted the
financial resources of an orphan child.) Much of the complexity in
Title III is a result of the multiplicity of goals the law is trying
to achieve; these include emergency planning, emergency response,
right to know, and an inventory of environmental releases of toxic
chemicals.

At least in concept, Title III represents an important departure
from other environmental legislation. Heretofore, Congress has
required the regulatory agency (usually the Environmental Protection
Agency) to develop standards that will limit emissions. Title III,
however, focuses on provision of previously unavailable data that can
improve decisionmaking about hazardous chemicals in the community.
For example, emergency planning will be aided by required lists of
"Extremely Hazardous Substances" (primarily airborne toxics) that
facilities store or use. More detailed chemical inventories also
assist planners as well as emergency responders, who may have access
to drawings showing relatively precise locations of specific chemicals
within each facility.

The law gives the right to all citizens, not just emergency planners or responders, to have access to these data. Based on informa-

tion about the chemicals stored, used, or emitted in their communities, people can decide whether these chemicals pose an unacceptable
risk and can work with facilities to reduce these risks -- perhaps
through changed storage methods, through reduction in quantities
stored on site, or through waste reduction programs. Title III
actually addresses only one of the impediments citizens have faced in
trying to reduce risks posed by hazardous chemicals in their communities: it makes the data available. With a few exceptions, the law
does not mitigate the considerable burden citizens must bear in
analyzing, understanding, and acting upon the data. After passage of
Title III, EPA worked with NOAA to modify CAMEO to address the emergency planning provisions of the statute(1). CAMEO II is implemented
in HyperCard for the Apple Macintosh family of micro-computers.

CAMEO II's Presentation Mechanism. The nature of HyperCard is essential to the effectiveness of CAMEO II and the Citizen's Helper.
Hypercard allows users to choose among many different paths through
a collection of information that may include graphics and fairly
sophisticated computations as well as standard text.
    A HyperCard program is called a "stack". Stacks are composed
of "cards" -- a card occupies the HyperCard window on the computer's
display; cards containing similar kinds of information have the same
appearance, or "background": this presents the user an important cue
concerning the relationships between parts of a stack.   Each card may
have its own graphics, fields for text, and buttons which, when
clicked, carry out commands ("scripts") associated with them by the
stack designer.   The presence of backgrounds, cards, fields, and
buttons are HyperCard's embodiment of the object-oriented programming
paradigm. A text field may be locked against modification, in which
case it is used for presentation of information only; if the field is
unlocked, it can be given a script that sends a message when the field
is modified -- thus a series of actions may be triggered by the user's
entry of data into a field. A button can be given a name (which may
or may not be displayed) and a distinctive graphic icon; although a
button's primary function is to invite the user to click on it with
the mouse, it may be sent an activating message by a script.
    A user's interaction with a HyperCard stack is reported inter-
nally by passing a message -- a button has been clicked, a field has
been changed through text entry. Thus HyperCard provides a robust
assortment of methods for soliciting input from the user, ranging from
transitions between cards to filling in specific textual information.
This feature proved very useful in recording the chemicals, in some
cases numbering in the hundreds, that an individual facility might
store. Once facility information has been entered, it may be passed
to subsequent cards containing information about different chemicals
without additional typing. If a change -- for example, in the name
or telephone number of the emergency contact person at the facility -
- is made on only one of the cards, however, it is passed to all the
others.

CAMEO II's Components. CAMEO II comprises four groups of stacks which
contain: 1) information about chemicals; 2) information about facilities which have reported storing, using, or manufacturing hazardous
chemicals; 3) aids for emergency planning; and 4) a separate air-plume

modeling program that can be employed from within CAMEO stacks. The following description is intentionally terse.

The primary constituents of the chemical-information group are Codebreaker, an extensive chemical name synonym-alias database, and a separate response information database that consumes more than seven megabytes of hard-disk storage. Navigating within either of these stacks can be quite slow, even though the transition from a selected chemical's card in the synonym database to the related response information is very rapid by virtue of special hypertext linkages. On a response-information card, the user may choose to see eight kinds of information, including physical properties, health effects, appropriate special fire-fighting equipment, and techniques for responding to other emergencies.

The group of stacks concerning facilities is capable of a good deal of depth of reporting. There is a stack for facility identification. In the stack for reporting materials covered under section 312 of Title III (manufacture, storage, or use), a separate card is used to tell about each reportable substance at a facility; as noted, common elements, such as the facility's name, address, and the emergency coordinator's name and telephone number(s) are entered automatically as cards for additional chemicals are added. Emergency planners may create links to sketches, created in another stack, identifying specific places within the facility where a substance is stored or used.

The group of stacks for emergency planning offers the following aids: a guide to formulating a plan; an outline for a plan under development; a repository for lists of people to contact; and a summary risk assessment for a chemical at a particular facility based on the information entered in other stacks. Planning a response to an emergency involving a chemical at an industrial site can be assisted by associating maps of the locality with the cards that describe the techniques to be used. The chemical plume-modeling program may be used to guide plans for evacuation in the event of an emission of a hazardous chemical.

CAMEO II's Distribution. More than three thousand copies of CAMEO II (herinafter referred to as CAMEO) have been distributed to fire and hazardous materials units in jurisdictions thoroughout the United States. Our sense is that while people are aware of the great benefits the program offers for emergency planning and fulfilling various requirements of Title III, wider use is inhibited at least in part by the fact that many emergency responders, including the Federal Emergency Management Administration (FEMA), are more accustomed to working with PCs than with Macintosh computers. The relative slowness of the older version of CAMEO also discouraged some users. Nevertheless, the program remains the single most comprehensive and easily-used software for managing the mountains of data submitted under Title III and for using that data both to create a plan and to assist in responding to emergencies.

The Citizen's Helper

The previous discussion has suggested that CAMEO strongly emphasizes two of the goals of Title III -- emergency planning and emergency response -- although the information they rely upon is also available under the law to the wider community. (CAMEO was not designed to be used for the emissions inventory, although there is no real reason it could not be so adapted. In the remainder of this discussion, however, we consider only the chemical lists and inventories provided under the sections of Title III not concerned with the emissions inventory (i.e., sections 302, 304, 311, and 312.) Indeed, CAMEO's primary gesture towards public participation is a stack that records citizens' requests for information.

This is particularly ironic in light of the fact that Hypercard has many features that make it especially appropriate for public access, including its connectivity and potential ease of use. Unfortunately, the relationship between the image of some of CAMEO's icons and the action performed by the stacks they represent is so abstract that they hinder, rather than help, the new user in navigating purposefully through the system of stacks.

In short, CAMEO contains information that is potentially useful for citizens making right to know requests. The information is not accessible, however, not least because CAMEO's intended audience was quite different. We decided to create a supplementary stack that was especially intended for citizen use, the Citizen's Helper Stack. The relationship of this stack and its adjunct, the Info Gatherer, to the CAMEO system is illustrated in Figure 1. The Info Gatherer is invoked, as needed, to collect lists of facilities that have reported storage or use of regulated chemicals, and of the chemicals reported. The Citizen's Helper uses these lists to aid its search for further information in the CAMEO stacks. Whereas emergency response planners or emergency responders must be familiar with the structure of the CAMEO group of stacks, a citizen seeking information should be able to obtain it without knowing about the existence of CAMEO.

As noted, Hypercard's connectivity allows users to move readily between related cards; CAMEO's large size and complexity means that sometimes this movement is slow. The Citizen's Helper is designed to minimize user frustration arising from slow transitions by anticipating the kinds of questions citizens are most likely to ask, by incorporating special search techniques to obtain the desired answers, and by creating efficient, direct connections that will speed transitions from one card to another. Similarly, icons were designed not only to be suggestive of the kinds of information or activity of the button, but also to be responsive to known citizen concerns.

Citizens' Concerns. Broadly speaking, the concerns a citizen might have concerning hazardous chemicals in the community are narrowly focused: "Will they affect my life(style)?" "Will they affect the value of my property?" "How much 1-2-dimethyl death is in my neighborhood?" "What are its effects on my health?"(2) Help in answering these questions can be provided by a stack that gives citizens intelligently focused, rapid access to CAMEO's information.

The goals set for the Citizen's Helper were flexibility in the format of requests for information, depth in pursuit of information, speed in carrying out the searches, and a truly friendly interface, including animated help for new users. Flexibility is embedded by letting the user choose among several levels of detail for each request, e.g., by allowing, where appropriate, restriction of the search to a ZIP Code or a range of ZIP Codes, by providing several primary topics for requests, and by accommodating changes in the

user's focus on the information uncovered.   As illustrated in Figure
2, the user may initially ask about:

o   facilities in the locality

o   a particular facility

o   chemicals reported in the locality

o   the presence of a particular chemical

o   the amount of a substance stored in the locality

o   the presence of chemicals that pose a particular threat to
        health

Each of these requests is effected by clicking on clearly-marked buttons.   If the desired information pertains to a "universe" (all the
facilities in the region, all the chemicals reported, etc.), it is
retrieved from lists created by off-line searches of the relevant
CAMEO stacks.   In other cases, efficient search techniques are used
to retrieve the information rapidly.   Subsequently, the user

o   having obtained a list of facilities, may select one of
        them and

o   ask about chemicals stored or used

o   ask for the name of the emergency response person

o   having obtained a list of chemicals, may select one of them
        and

o   ask about the weight stored or used

o   look at the health-effects information in CAMEO

"Selecting" refers here to highlighting a name in a text field on the
card.   "Asking" is accomplished by clicking a button.   An initial
request based on a health effect produces a list of chemicals reported
that are associated with the health effect, and a matching list, for
each chemical, of the facilities at which the substance is stored or
used.   Either of the above lines of further inquiry is available in
this case.   Figure 3 illustrates the display that results from requesting a list of facilities, then the facility information concerning one of them.   Figure 4 shows the result of a request for a list
of toxic chemicals stored or used in the region.

     The overall effect is to allow a user to pursue his/her interests
as they arise.   A list of chemicals stored at a nearby facility might
prompt either a request for a list of other facilities storing a
particular chemical or a request for information about the health
effects of one of the chemicals.   Changes in the user's interests are
to be expected as an information-gathering session progresses.   The
Citizen's Helper stack anticipates this need for flexibility by

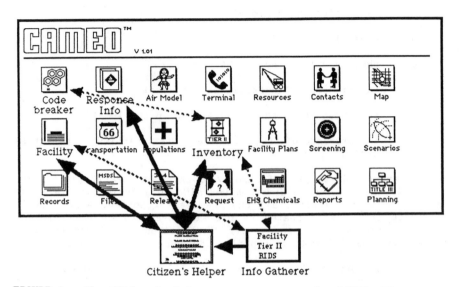

FIGURE 1. The Citizen's Helper as gateway to the CAMEO II system. The arrows represent the retrieval of information from CAMEO for display in the Citizen's Helper stack.

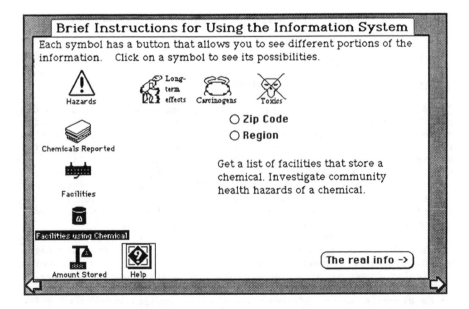

FIGURE 2. A sample of the Citizen's Helper stack's introduction to its buttons and the information they elicit.

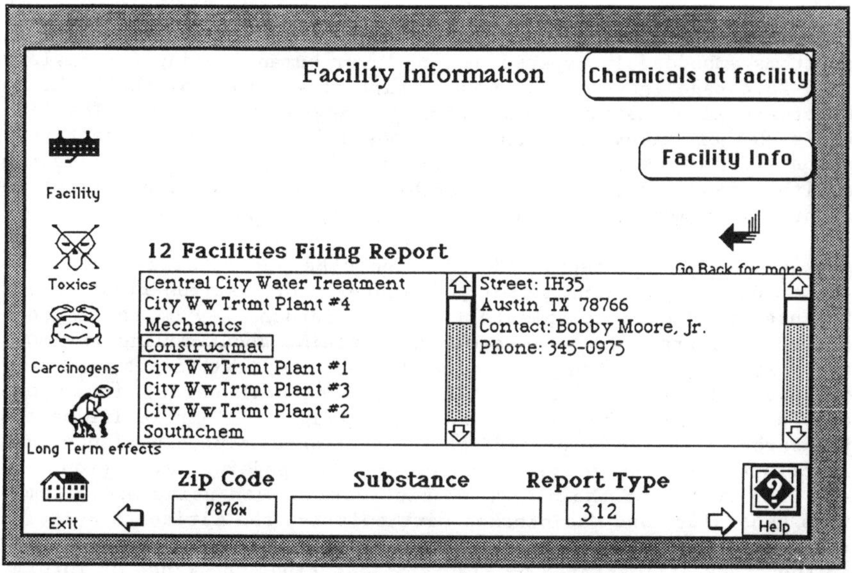

FIGURE 3. The flexible information display mechanism in the Citizen's Helper stack. The user first requested the list of facilities in the ZIP code shown, and then the facility information, shown at the right, for the facility shown highlighted.

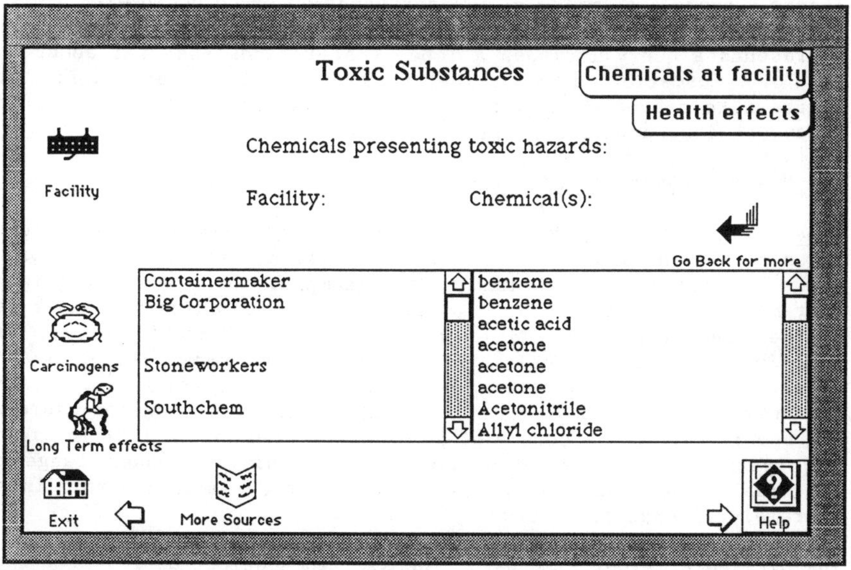

FIGURE 4. A portion of the response to a request for a list of toxic chemicals used or stored in a hypothetical geographic region.

including buttons for other queries on each of the cards on which the results of a search are reported. Speed in carrying out the searches is a serious issue, given the size of some of the CAMEO stacks, in spite of the efficient search algorithms embedded in HyperCard: a clever human looking for "xylene" in a 2674-page encyclopedia would start at the back of the book, but Hypercard always starts from the beginning of the stack. The Citizen's Helper makes use, wherever possible, of unique identifying numbers for cards in stacks. We created index tables of chemicals by health effect; these make it possible to find the health information about xylene more rapidly than even the clever human could.

Access for All Citizens. The Citizen's Helper stack also tries to accommodate the most naive micro-computer user. At the first screen, the user need only be able to find the 'return' key on the keyboard in order to transfer to a card that explains about moving the mouse and clicking its button, working with scrolling text fields, and other features of HyperCard. For the first several cards, it is only necessary to move the mouse's on-screen image to a particular part of the screen in order to proceed to the next card. The experienced user is offered alternative ways of achieving purposeful navigation.

Regardless of the user's sophistication with using micro-computers generally, and Macintoshes particularly, the Citizen's Helper is designed to assist the user's quest for information. The second card in the stack, illustrated by Figure 5, describes the kinds of information that can be obtained via the Citizen's Helper. The third card, shown in Figure 2, shows the most salient button icons: clicking on each icon triggers a display describing the type of information that can be elicited through its use in an actual request for information. In this way, all users have a guard against suprises when they initiate a query. (Similarly, all databases ought to begin with a brief description of the contents for new users.) Each icon button that represents a query includes a label that reminds the user about the type of information to be obtained through use of the button. In stark contrast to database programs with "command line" interfaces, it is always obvious how to exit from a session.

The Use of Expertise and the Citizen's Helper's Future

Expertise for Laymen. Most intelligent advisory systems are designed for a particular audience -- usually a professional audience -- or to answer a particular question. For example, there are intelligent systems to assist facilities in estimating their emissions under yet another requirement of Title III; other such expert systems are described in this symposium. As far as we know, the Citizen's Helper is unique in addressing a lay audience. There is a difficult tradeoff in designing a system for such an audience, however; some of the specificity and accuracy of an expert system may be lost in making the system useful for an unknown but certainly broader range of concerns as well as in deploying some of the system's power in an extremely friendly interface.

The Citizen's Helper's Future. The Citizen's Helper has evolved from a stack that demonstrated the feasibility of various search and retrieval functions, to one in which many stable indexed lists of

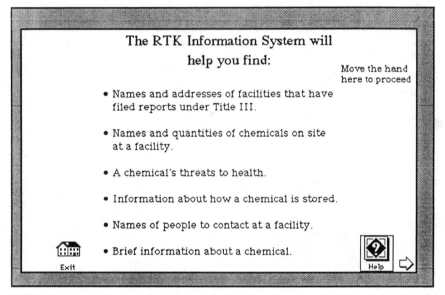

FIGURE 5. A guide to the kinds of information available with the aid of the Citizen's Helper stack.

locations for information are used to reduce the length of an interested user's interaction. The maintenance functions -- creating
annual lists of facilities that filed reports and of the substances
reported -- are now separate from the "presentation" aspects of
the system. Such a separation makes the user's interaction more
satisfying because of its brevity.

The Info Gatherer, which creates the lists of substances reported
and establishes the linkages to the CAMEO chemical-information database using CAMEO's information base of chemical synonyms, is intended
to be exercised by someone with experience with CAMEO and with chemicals. This process provides the opportunity to set up meaningful
relationships between problematical substances, such as those reported
by trade name or mixtures, and the CAMEO information about chemicals.

Some users may want to see a display from CAMEO's air dispersion
model, which requires additional input. Guidance as to reasonable
input for the dispersion model would be very beneficial to emergency
planners as well as to interested citizens, and would help ensure that
results are meaningful. Assistance in interpreting the plume display
would also be helpful to citizens.

Another enhancement would be to include a mechanism whereby a
user who is not satisfied with the results of a session could post a
request for information inaccessible through the Citizen's Helper.
(It is anticipated that the Citizen's Helper would be installed at a
public place, such as the town library.) These requests would be
answered by a local expert, e.g., a member of the LEPC designated to
monitor use of the Citizen's Helper. Some LEPCs have already discovered the value of providing mechanisms through which citizens may
arrange to receive a call from appropriate experts. (The Columbus,
Ohio, LEPC has provided such opportunities through its printed materials with a mail-in coupon; other LEPCs have used similar mechanisms.)
An analysis of patterns in such requests could form the basis for
further refinements of the intelligent information retrieval system.

The Citizen's Helper could also serve as the vehicle for much
needed research on the kinds of information citizens want from the
Title III data, and on the effects that the mode of presentation has
on their perception of risk or ability to ameliorate a hypothetical
(or real) situation. The data concerning citizens' information needs
could be elicited simply by adding scripts that recorded the actions
taken by information seekers. The effects of mode of presentation
would be obtained by devising several forms of delivering the same
information.

Title III gave citizens the right to know about hazardous materials in their communities, but also placed on them the largest part of
the burden for obtaining, understanding, assessing, and using the
data. Even the most willing environmental and emergency officials
at the state and local levels are hindered by Congressional failure
to appropriate funds for implementing Title III. Offices that are
already using CAMEO, however, can reduce the burden on the staff to
answer right to know requests and especially requests for supplementary data interpretation by using the Citizen's Helper stack. In the
best of all worlds, a computer would be available in the public
library for citizens to use, following interesting leads in the data
as they choose.

As citizens increasingly demand access to environmental data of
all kinds, it will be increasingly important to ensure that they

understand and can exploit the connections among the data, that they can obtain reasonable interpretations of it, and that they do not become frustrated and angry because data are too difficult to acquire. The Citizen's Helper stack constitutes an effort to enhance an intelligent information system directed toward the needs of environmental and emergency response professionals, making the system responsive to the needs of citizens who will be affected by the decisions of those professionals. With additional modules and expertise built in, the system can become an even more effective tool for public education and empowerment.

Acknowledgments

Development of the Citizen's Helper was funded in part by the Bauman Foundation. Copies of this program, which can be readily used by anyone who owns CAMEO II, are available from the author at cost.

Literature Cited

1.   CAMEO II Users Manual, National Oceanic and Atmospheric Administration, National Safety Council, and Environmental Protection Agency, 1988, p xvi.
2.   Hadden, S. G. A Citizen's Right to Know: Risk Communication and Public Policy; Westview Press: Boulder, CO, 1989; pp 13-14.

RECEIVED March 19, 1990

Chapter 9

An Expert System To Diagnose Performance Limiting Factors at Publicly Owned Treatment Works

Linda Berkman, Mark Lennon, and Keith Law

Eastern Research Group, 6 Wittmore Street, Arlington, MA 02174

This paper introduces POTW EXPERT, a diagnostic expert system based on an economic, non construction-oriented approach for optimizing Publicly Owned Treatment Plant (POTW) performance. Approximately two-thirds of the nation's operating POTWs have effluent quality or public health problems; it has been estimated that $36.2 billion would be required to address these problems. POTW EXPERT, a joint effort implemented by the U.S. Environmental Protection Agency and Eastern Research Group, Inc., assists evaluators in recognizing plant performance problems by modeling the actual evaluation strategy employed by experts concerned with diagnosing POTWs. By observing symptoms and using experiential rules of thumb, POTW EXPERT identifies factors which impede the optimization of plant performance. This paper discusses our knowledge engineering and elicitation approach, the formulation of heuristically guided diagnosis procedures, and our methods for qualifying information and dealing with suspect or marginal data. Validation of the system will be examined both in terms of how the system should be validated and who should perform the validation.

Wastewater is the flow of used water from a community. Wastewater comes from five sources: residential, commercial, and industrial sources, and storm and ground water. Treatment of wastewater is concerned with the removal of constituents which: will deplete oxygen resources of the receiving waters to which they are discharged; may stimulate undesirable growth of plants and organisms in the receiving water; or will have adverse health or other effects on downstream water uses. In the United States, the need for better performance from existing wastewater treatment facilities is extensive.

In 1986, the U.S. Environmental Protection Agency (EPA) reported that 10,131 of 15,438 operating Publicly Owned Treatment Works (POTWs) have documented effluent quality or public health problems. EPA

0097–6156/90/0431–0120$06.00/0

estimated that $36.2 billion would be required to address these problems (1). Federal funding to support POTW improvements has been decreasing, however, as the federal government has asked states and local governments to shoulder an increasing share of the financial responsibility for maintaining and upgrading POTW performance. A clear need among POTW regulators and operators is to optimize the performance of existing facilities before implementing costly new design modification and construction projects.

A number of engineering firms and other organizations have developed methodologies (including several software products) to assist in diagnosing the causes of POTW performance problems. One of the most successful of these methodologies is the Comprehensive Performance Evaluation/Composite Correction Program (CPE/CCP) approach developed by Process Applications, Inc., of Fort Collins, Colorado. Based on nearly a decade of EPA- and state-sponsored investigation, the CPE/CCP approach represents a structured methodology (1) to evaluate the capacity of existing POTW processes to meet effluent permit requirements and (2) to identify cost-effective, non construction-oriented options to improve plant performance. Process Applications' methodology has been documented in EPA's *Handbook, Improving POTW Performance Using the Composite Corrective Action Approach* (2).

Based on the proven track record and widespread applicability of the CPE/CCP approach, EPA's Office of Research and Development/Center for Environmental Research Information decided in 1988 to fund development of an expert system to incorporate the expertise embodied in the CPE/CCP methodology. This report describes the concept, design, and implementation of this expert system. The system, POTW EXPERT, was beta-tested in September, 1989, and is currently in limited-distribution testing by EPA.

The CPE/CCP Approach to POTW Evaluation

The CPE/CCP approach to POTW evaluation proceeds in two distinct but linked phases. The Comprehensive Performance Evaluation identifies and prioritizes the design, operational, maintenance, and administrative factors which may be the cause(s) of suboptimal POTW performance. The Composite Correction Program implements procedures (e.g., changes in plant operations, training for plant operators) to address these factors. The CPE and CCP are linked in that implementation of the CCP may feed back to affect the content or order of the prioritized list of factors affecting plant performance. The distinction between CPE and CCP marks a logical problem boundary for an expert system built around the CPE/CCP approach; development of POTW EXPERT to date has focused on the Comprehensive Performance Evaluation.

A Comprehensive Performance Evaluation encompasses two primary sets of activities:

1.   Evaluation of Major Unit Processes
2.   Identification and Prioritization of Performance Limiting Factors

Evaluation of Major Unit Processes. The goal of this evaluation is to determine whether existing biological and chemical treatment processes are adequate to meet effluent quality goals, given existing and projected influent volumes and organic loading. The evaluation is based on analysis of data and on-site observations made for each major unit process in the plant, and assignment of points based on process design and capacity. The end result of the evaluation is a categorization of the plant into one of three types:

Type 1   Capability of major unit processes does not limit plant performance. Performance problems are potentially related to plant operation, maintenance, or administration, or to unit process malfunctions which can be corrected with only minor facility modifications.

Type 2   Capability of one or more major unit processes is marginal. Efforts to optimize all aspects affecting the unit's capability are warranted prior to upgrade of the unit process.

Type 3   One or more major unit processes is inadequate to satisfactorily treat existing influent volumes or organic loads. Plant performance cannot be expected to improve significantly until the limiting process(es) are upgraded.

Analysis of major unit process capacity allows POTW owner/operators to pinpoint many design-related causes of plant performance problems. Categorization of a plant as Type 1 or Type 2 suggests that the capability exists and performance can be significantly upgraded without a major construction program, and that degraded plant performance can be related to problems in plant operation, maintenance, and/or administration which can be addressed with relatively low-cost solutions.

Identification and Prioritization of Performance Limiting Factors. Once the plant has been scored, a Comprehensive Performance Evaluation proceeds to identify and prioritize the factors that can be correlated with plant performance problems. With a database of hundreds of plant analyses completed, Process Applications has identified a set of 66 factors -- related to plant administration, maintenance, design, and operation -- which may cause degraded plant performance. The real expertise developed by Process Applications consists of the engineering and socio-economic knowledge and judgment required to sift through literally hundreds of observations and data points related to plant design and performance, and to glean information from interactions with plant operators and administrators, in order to identify those few factors which are the primary causes of plant performance problems.

The Performance Limiting Factors (PLFs) are structured as a series of questions which require the evaluator to judge the extent to which each PLF may be related to an identified performance problem. An initial analysis eliminates most PLFs from consideration as causes of poor plant performance, and leaves a subset (usually no more than 10 to 15 PLFs) which contribute directly to the identified performance problem. The evaluator then prioritizes these remaining factors. The end result of the process is a prioritized list of PLFs which defines those areas of plant operation, administration, design, and/or maintenance which should be addressed first in attempting to upgrade plant performance. The CPE/CCP methodology recognizes that correction of one PLF may result in the identification of additional PLFs, not included on the original prioritized list, which contribute to degraded performance but which were masked by PLFs on the original list. The CPE/CCP is therefore an iterative process -- the list of PLFs should be re-analyzed and re-ordered as improvements are made to the factors first identified as the most likely causes of unacceptable plant performance.

The Expert System

The objective of the POTW EXPERT development effort is to allow users to rapidly and effectively evaluate the cause(s) of degraded performance at POTWs that fail to meet effluent quality targets. The system achieves this objective by making the proven diagnostic expertise embodied in the CPE/CCP approach available to the community of wastewater treatment regulators and operators responsible for achieving and maintaining POTW effluent quality.

POTW EXPERT address three types of secondary wastewater treatment processes: suspended growth, fixed film, and stabilization ponds. The expert system emphasizes meeting National Pollutant Discharge Elimination Standards (NPDES) for secondary treatment facilities (30 mg/l BOD_5 and TSS) with a maximum average daily flow of 20 million gallons per day (mgd).

Hardware and Software Requirements and Selection.   A number of requirements influenced the early design phases for POTW EXPERT. These were related primarily to the user community for the system, development and delivery hardware, and the need to continually update the system.

Users   POTW EXPERT is targeted to a user community of potentially thousands of POTW owner/operators, state and local regulators, and consulting engineers, among whom the development team could assume only rudimentary computer knowledge. In addition, a fundamental premise of the development effort has been that the system would be provided to users at little or no cost; the size of the user community therefore precluded the selection of commercial expert system software with significant run-time, distribution, or hardware costs.

Platform   Because of the nature of the intended user community, POTW EXPERT has been developed for delivery on PC/XT class machines with no more than 640K of memory. (Better performance is obtained, however, if the system is installed on an AT class machine.) Because of the very large knowledge base we anticipated for the system, we attempted to identify an expert system shell with an efficient paging mechanism to allow effective use of memory and the storage capacity of a hard disk.

Updates   Because the knowledge base incorporated into the CPE process is continually being expanded and refined, and because POTW performance requirements are the subject of active regulatory intervention, we faced the demand to construct a very robust system -- robust both in allowing straightforward expansion to the knowledge base and inference mechanism, and in assuring that system performance would not become unpredictable as changes and updates are incorporated.

During project initiation, we evaluated over fifteen candidate expert system shells for suitability for the proposed system. The majority of shells operating in our required delivery environment are simple, rule-based systems. Many of these tools are very simple to learn and use, offering appealing developer and user interfaces. As such, they are attractive for limited applications or for relatively inexperienced developers. However, a problem as large and complex as that defined by POTW EXPERT cannot be force-fit into the single processing paradigm offered by most PC-based shells. The nature of the POTW EXPERT's diagnostic/classification is best suited to a hybrid knowledge representation structure, coupled with flexible information processing capabilities. High end shells available for the PC (e.g., Goldworks, Nexpert/Object, KEE) offer powerful implementations of a hybrid system, but either demanded significant enhancements to the PC delivery platform or entailed prohibitive run-time charges for the large POTW EXPERT user community.

We ultimately selected the shell ALEX, developed by Harris and Hall Associates (P.O. Box 1900, Port Angeles, WA 98362), as the development vehicle for POTW EXPERT. Developed specifically to support diagnostic procedures for complex engineered systems, ALEX is a collection of classes and methods which allows programmers to develop object-oriented expert systems. The shell employs a structure of active demons and changing levels of belief or probability to simultaneously monitor a large number of factors which may contribute to degraded performance of the engineered system under analysis. Written in Smalltalk/V, a PC-based implementation of Xerox's object-oriented Smalltalk language, ALEX met the exacting requirements imposed on hardware and software selection:

• Window- and menu-driven, ALEX's user interface is clear, simple, and concise. ALEX may be run with or without a Mouse.

• ALEX and Smalltalk/V create a virtual memory structure with an efficient paging facility to minimize the memory limitations of the PC delivery platform.

• ALEX allows expert systems to be structured in completely independent modules. Variables, once defined, can be shared by a number of modules. The expert system can thus be developed in stages, and system behavior remains predictable as the system is modified and expanded.

• Inferencing is controlled by the level of belief or probability attached to objects ("demons," defined below), which can be examined at any point during system development or consultation. This inference mechanism allows the developer to retain straightforward and precise control over inferencing.

• The development platform is totally open. Specifically, both the ALEX and Smalltalk/V source code are available to the developer. This feature has allowed us to develop a number of in-house enhancements both to fine-tune ALEX as a development tool for POTW EXPERT, and to improve the performance of ALEX and Smalltalk/V in the PC/XT delivery environment.

Knowledge Engineering

Because POTW EXPERT was based on the existing EPA/Process Applications *Handbook*, the domain experts for the system were pre-selected. Also, much of the knowledge to be incorporated in POTW EXPERT was codified in the *Handbook* prior to initiation of system development; specifically, many of the procedural calculations required for the major unit process evaluation were specified in the *Handbook*, and the structure and content of the Performance Limiting Factors were outlined. For these reasons, functional specifications and much of the early system design could be completed before formal knowledge engineering was initiated. Formal knowledge engineering with the Process Applications experts was also simplified, both because much of the required knowledge had already been structured and recorded, and because we were able to produce limited function prototypes of POTW EXPERT for demonstration to the domain experts very early in the knowledge engineering process.

Nonetheless, knowledge engineering for POTW EXPERT was complex. The major knowledge engineering effort has been focused on the Performance Limiting Factors. Most of the sophisticated expertise required to complete a CPE is related to identification and ranking of the PLFs, but the *Handbook* provides little of the information required to complete the PLF evaluation. The *Handbook* includes some 600 data points that may contribute to assessment of the PLFs, and the structure of the PLFs themselves provides some guidance on the data most important to assessing each factor. But, in general, the *Handbook* does not provide sufficient information to make the sophisticated judgments required to identify and prioritize the PLFs most directly related to poor POTW performance.

The major unit process evaluation also demanded a significant knowledge engineering effort. During implementation of this largely procedural module of POTW EXPERT, we encountered many instances in which complex engineering judgment was required to assess process capacity. These situations -- related primarily to assessment of oxygen transfer capacity, estimation of sludge volumes, and evaluation of sludge handling capacity -- were rarely apparent in the calculations presented in the *Handbook*, and required extensive interaction with the domain experts to elicit the heuristics crucial to their resolution.

Knowledge engineering for the system was complicated by the fact that the domain experts were located 2,000 miles from the development team, and so were not accessible for routine, face-to-face communications. For this reason, we employed a knowledge engineering approach centered on three critical stratagems: (1) we devoted more effort than is typically required to developing domain expertise among our design/development team; (2) we completed as much prototype design and development as possible on the basis of the *Handbook* before initiating on-site knowledge engineering; and (3) we devoted significant effort to preparing detailed but uncomplicated paper models of the major unit process and PLF heuristics for review by the domain experts.

On-site knowledge engineering for the initial prototype was completed in two three- to four-day sessions between the domain experts and two members of the design/development team. These sessions took the form of structured interviews; the structure for the interviews was determined jointly by the organization of the EPA/Process Applications *Handbook* and by the design of the prototype POTW EXPERT. On-site knowledge engineering was supplemented by frequent telephone and mail communications before and after the two knowledge engineering sessions. Virtually all of the significant elements of reasoning incorporated into POTW EXPERT were confirmed by a formal review of the PLF and major unit process paper models developed as part of our knowledge engineering strategy.

Structure of POTW EXPERT

POTW EXPERT's top-level architecture consists of four modules (Figure 1). These are:

1. Data Entry
2. Major Unit Process Evaluation
3. Evaluation of Performance Limiting Factors
4. Report Generation

Data Entry. The Data Entry module requests information on specific physical plant characteristics and on the characteristics of plant influent. Information collected in this module defines the data collected and processing operations performed in subsequent modules. Data entry screens are generated dynamically based on the specific plant characteristics defined in the initial data entry screens; they are tailored specifically to the design and equipment characteristics for the POTW under analysis (Figure 2). Although most POTWs have a similar overall layout (e.g., primary clarifier, aerator, secondary clarifier, sludge handling system), there is tremendous variation in the particular equipment employed and in the specifics of process configuration. For example, clarifiers differ in size, shape, capacity, means of sludge removal, and other variables; aerators differ in the aeration process employed, oxygen transfer capacity, oxygen transfer efficiency, and other variables. POTW EXPERT has been developed to be useful across a broad spectrum of the POTW community, and therefore is able to analyze the capability and operation of a large number of combinations of specific unit processes. The first two data entry screens request information on specific process characteristics of the POTW under analysis, and use this information to set branching options that control data entry requests and

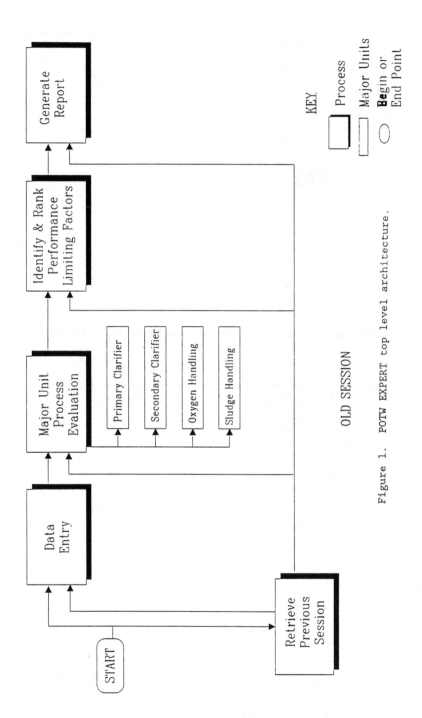

Figure 1. POTW EXPERT top level architecture.

CLARIFIER – CIRCULAR

```
Greater Flow for Clarifier (see detail)
Configuration
Sludge Removal Mechanism
Surface Area
Depth near the Weirs
Minimum Return Sludge Flow Possible
Maximum Return Sludge Flow Possible
Is there a rag problem?
Is There a Grit Build-up Problem?
Secondary Clarifier Blanket Depth Testing
RAS Flow Rate Testing
Mass Control Testing
Settling Rate Testing
```

ESC: Quit Data Entry   F3: Text   F8: Clear Field   F10: Menu   Enter: Accept

CLARIFIER – RECTANGULAR

```
Greater Flow for Clarifier (see detail)
Is the Launder at the very end?
Surface Area of Launders
Return Sludge Removal Mechanism
Surface Area
Depth near the Weirs
Minimum Return Sludge Flow Possible
Maximum Return Sludge Flow Possible
Is there a rag problem?
Is There a Grit Build-up Problem?
Secondary Clarifier Blanket Depth Testing
RAS Flow Rate Testing
Mass Control Testing
Settling Rate Testing
```

ESC: Quit Data Entry   F3: Text   F8: Clear Field   F10: Menu   Enter: Accept

Figure 2.   POTW EXPERT data entry screens tailored to the specific characteristics of the POTW under analysis.

information processing in the subsequent Data Entry and Major Unit Process Evaluation Modules.

Because of the great variation in the details of POTW design and equipment, use of standardized data entry screens might have proved confusing to users, and would have resulted in the collection of superfluous data (both slowing system operation and taxing the memory limitations of our delivery platform). The object-oriented knowledge representation structure made possible by ALEX and Smalltalk have allowed us to implement a very flexible data entry system matched specifically to the individual POTW under analysis. An "explain" facility is available for all data requested, and default values may be used if a user is unable to provide selected plant information.

Users need not complete data input in the Data Entry module. Any information omitted by the user will be requested interactively when required by POTW EXPERT during the Major Unit Process Evaluation or Evaluation of Performance Limiting Factors. POTW EXPERT will attempt to reason in the absence of user input data, but will not conclude its reasoning process if critical input is missing.

Major Unit Process Evaluation. Along with the Evaluation of Performance Limiting Factors, the Major Unit Process Evaluation forms the heart of POTW EXPERT. This module evaluates the capability of the POTW's engineered and biological processes to handle plant influent loads and meet effluent permit requirements. Processes evaluated include: primary clarifier, aeration system, secondary clarifier, and sludge handling system (Figure 1). This module offers the capability to analyze a large number of combinations of specific plant configuration and process design and equipment.

Much of this module is procedural, but it uses expert systems reasoning in a number of areas. One of the most common causes of degraded POTW performance identified by Process Applications is the misunderstanding or misinterpretation of plant operations and/or performance characteristics on the part of POTW operators. For this reason, some of the most difficult judgments made during the course of a CPE involve comparing reported plant characteristics and performance against estimates based on an engineering evaluation of plant design. POTW EXPERT mimics this reasoning process in the Major Unit Process Evaluation; particularly critical areas are the aeration system and the sludge handling system. If POTW EXPERT determines that user-supplied information about these processes is suspect, the system will request the user to confirm the information. If information so confirmed remains questionable, the system uses its internal engineering-based values to estimate major unit process capacity, and flags the operator's understanding and or application of POTW processes as a probable factor contributing to degraded plant performance.

The primary output of this module is a categorization of the POTW according to its capability to produce effluent of required quality. The module also reports a comparison of process capacity against plant influent loads, and indicates which process(es) may be limiting the plant's ability to meet effluent quality goals. Many of the calculated results generated by this module also serve as input to the Evaluation of Performance Limiting Factors.

Evaluation of Performance Limiting Factors. The 66 PLFs identified by Process Applications fall into four categories: design; operation; maintenance; and administration. A few examples of Performance Limiting Factors are:

Type: Administration
Factor: Plant Administrators' Familiarity with Plant Needs

> Do the administrators have a first-hand knowledge of plant needs through plant visits, discussions with operators, etc.? If not, has this been a cause of poor plant performance and reliability through poor budget decisions, poor staff morale, poor operation and maintenance procedures, poor design decisions, etc.?

Type: Design
Factor: Unit Design Adequacy, Secondary Process Flexibility

> Does the unavailability of adequate valves, piping, etc. limit plant performance and reliability when other modes of operation of the existing plant can be utilized to improve performance (e.g., operate activated sludge plant in plug, step, or contact stabilization mode; operate RBCs in step loading mode)?

Type: Operational
Factor: Process Control Adjustments -- Application of Concepts and Testing to Process Control

> Is the staff deficient in the application of their knowledge of sewage treatment and interpretation of process control testing such that improper process control adjustments are made?

This module evaluates each of the 66 factors, and ranks them according to their impact on plant performance. Factors are classified into one of four categories:

Class A   Continuing major impact on plant performance

Class B   Continuing minor impact on plant performance, or significant impact on a periodic but infrequent basis

Class C   Minor, occasional impact on plant performance

NR   Factor not associated with degraded plant performance; no rating

Much of the information required to reason about the PLFs is carried from the Data Entry and Major Unit Process Evaluation modules. Using an essentially forward-chaining inference mechanism, the system uses information already available from these modules to complete a preliminary assessment of each PLF. The initial assessment collects and analyzes information to conclude whether a factor adversely affects plant performance. This evaluation has two possible end points:

a)   Cite the factor as a Performance Limiting Factor

b)   Cite the factor as "No Rating" (does not limit plant performance)

All factors that have been cited as PLFs are then further evaluated to assess the nature and severity of their impact on plant performance. Using information available within its knowledge base, POTW EXPERT recommends a classification (Class A, B, or C) for cited PLFs. In its current version, POTW EXPERT generates a recommended classification for approximately half of all PLFs (those related to plant design and operation); the user is requested to make this classification for the remaining cited PLFs (related to plant administration and maintenance). The user also has the option to manually override the system-generated classification of any PLF. "Explain" text is available for each cited factor to assist the user in making or reviewing a classification.

Output from this module consists of a ranked listing of the Performance Limiting Factors. POTW EXPERT does not rank PLFs within each Class, but generates a report listing PLFs by Class and providing the information that resulted in each classification.

Report Generation. Reports generated by POTW EXPERT include:

1.   Classification of the POTW according to its unit process capacity and capability to meet effluent quality goals. This report includes information on each major unit and the overall rating of the plant.

2.   Ranked listing of Performance Limiting Factors. This report groups PLFs by Class, and reports on the reasoning that resulted in each classification.

Knowledge Representation

POTW EXPERT's knowledge base is represented in two primary structures:

1.   In the Major Unit Process Evaluation module, information about plant design and operating characteristics is stored in a hierarchy of frame-like classes. The logic for gathering this information is controlled by a structure of ALEX demons and variables.

2.   In the Evaluation of Performance Limiting Factors module, knowledge about the PLFs is maintained by a structure of demons and variables coded in ALEX.

Frames. In designing and implementing the Major Unit Process Evaluation, we needed to address the fact that most POTWs share a common basic layout, but differ in a large number of design and process details. To efficiently request, handle, and process information about the major unit processes, we required a structure that recognized the basic similarities between all POTWs, and the information common to all major unit process analyses, but provided the flexibility to handle the large number of specific process types and design details represented in the POTW population.

Smalltalk's object-oriented system of classes allowed us to achieve these ends. Smalltalk's object-oriented paradigm allows developers to design systems that model human reasoning and language in terms of objects and actions taken on objects. As analytical thinkers, humans usually classify knowledge hierarchically, breaking problems down into more easily resolved subproblems. Smalltalk mimics this process by classifying information into a hierarchy of classes with related characteristics.

At each level in a hierarchy, a class stores information common to all classes deeper in the hierarchy (Figure 3). Succeeding levels store more specific information particular to a unique plant configuration or process type; at the bottom of the hierarchy are classes that store information related to the specific types of machinery operating or the specific details of process operations. For example, all POTW EXPERT consultations access a top-level class called "CPE," which holds very general information about the plant and the particular CPE being performed. The next level class is "SuspendedGrowthFacility," a subclass of "CPE," which handles information common to all suspended growth POTWs. Succeeding levels handle information on specific unit processes of the POTW and specific design and equipment characteristics of the plant.

This hierarchical class structure was also critical to implementation of an efficient data input system for POTW EXPERT, and to efficient processing throughout the system. Preliminary plant design information input by the user identifies to the system those frames which will be active during a system consultation. The data entry module then requests only information required to complete these frames. In turn, once a unique frame path has been defined for each major unit process during an analysis, the Major Unit Process Evaluation and Evaluation of Performance Limiting Factors need refer only to a top-level frame to access all general and specific information concerning that process.

Demons and Variables. ALEX provides a three-tiered knowledge representation structure of contexts, demons, and variables. Contexts are top-level structures that isolate individual major areas of the problem under analysis. Demons are the central knowledge representation structure; they are essentially frame-based information storage nodes with slots both to hold information on the content of the demon and to reflect the impact of one or more variables on the status of the demon. Variables may hold user-input, default, or calculated information.

CONTEXTS -- The larger problem of evaluating POTW Performance Limiting Factors is divided into contexts. Contexts define the skeleton of an expert system implemented in ALEX; they correspond to relatively isolated sections of the problem under investigation, in our case, to the major evaluation areas of the POTW performance analysis. A hierarchy of contexts can be defined, so that this top level architecture can be subdivided into a number of sublevels. In POTW EXPERT, the top level contexts are system managers, which

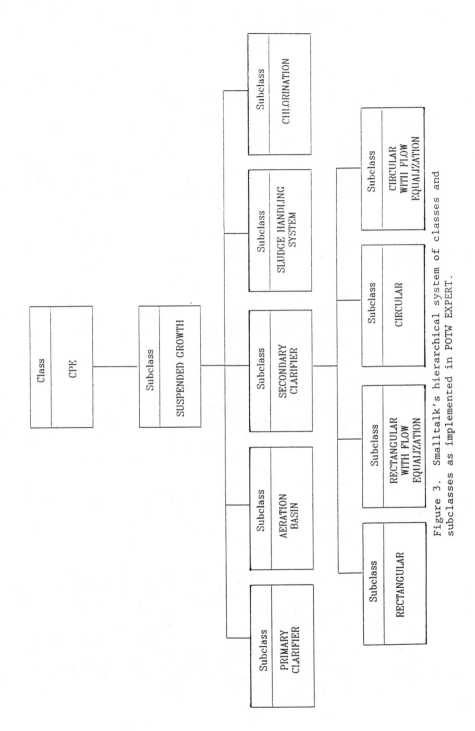

Figure 3. Smalltalk's hierarchical system of classes and subclasses as implemented in POTW EXPERT.

determine which of the most general system functions will be operated
during a session (e.g., data entry, major unit process evaluation,
identification and prioritization of PLFs, report generation). Below
these, a second level of contexts defines the four functional areas
examined during operation of POTW EXPERT -- aerator, secondary
clarifier, sludge handling, and Performance Limiting Factors. Each
of these contexts, in turn, may preside over a third level of
contexts, each defining a specific classification of information to
be examined during the course of a system consultation. For example,
the "Performance Limiting Factor" context controls investigation of
all PLFs related to the plant by managing a number of demons
(explained below), each assigned to a single PLF. Each of these
demons, in turn, transfers control to a PLF-specific context that
manages the investigation of a single PLF by responding to and acting
on information accumulated in additional demon(s) associated with the
context.

DEMONS -- Demons are the central knowledge representation
structure in an ALEX expert system and are one of the actual
repositories of knowledge. In the Smalltalk vernacular, the demon
is an object that has a fairly large number of instance variables.
A demon corresponds to each of the areas in which a cause of a POTW
performance problem may be isolated; in POTW EXPERT, at least one
demon has been defined for each of the 66 potential Performance
Limiting Factors identified by Process Applications.

Three instance variables associated with each demon are its
"initial belief," "current belief," and "minimum threshold"
(Figure 4). Initial belief is a developer-defined default. At the
beginning of an ALEX consultation, current belief equals initial
belief; current belief changes as variables related to the demon are
resolved. The minimum threshold defines the minimum level of current
belief that must be obtained in order for actions associated with the
demon to be processed.

Each demon accumulates evidence that its associated PLF is or is
not a factor that limits plant performance. This evidence is
represented by the demon's level of current belief, which is
determined by the effect of the individual variables related to the
demon. Many demons may "observe" the same variable, but each demon
will have a distinct reaction to the value of the variable. For
example, resolution of a variable which increases the level of belief
in one demon can simultaneously decrease the level of belief in one
or any number of other demons. A critical design component of an ALEX
expert system, therefore, is the definition of which variables will
be observed by each demon, and determination of the impact each
variable will have on the belief of each associated demon. Once a
demon acquires control of a system consultation, investigation of
variables related to the demon continues until all variables related
to that demon have been evaluated. At the end of the consultation,
the final level of belief of the demons also defines the system's
conclusions -- with POTW EXPERT, the PLF demons' final levels of
current belief determine whether or not each PLF is judged to be a
cause of degraded plant performance.

VARIABLES -- Individual pieces of information in an ALEX
knowledge base are stored in variables. Variables may be either
conditional (True, False, Unknown), multiple choice, or numerical.

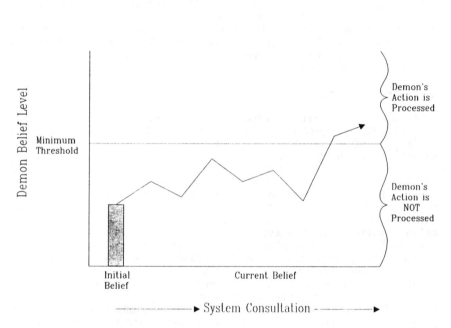

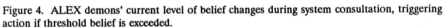

Figure 4. ALEX demons' current level of belief changes during system consultation, triggering action if threshold belief is exceeded.

Variables may be resolved either by using a default value, interactively requesting user input, by consulting a database, or computing a value from other variables. Variables are related to one or more demons. The developer defines which demons will watch each variable, and the impact which the resolved variable will have on each watching demon; for example, a variable resolved as "Filter Rarely Cleaned" might simultaneously increase the belief of a demon watching for problems in plant operation and decrease the belief of a demon watching for problems with filter capacity. A variable affects the level of belief of a given demon through a series of "weights." A different weight is attached to each possible resolution of a variable. A weight of one will leave a demon's belief level unaffected; a weight greater than one will increase a demon's belief, while a weight less than one will cause a demon's belief to decrease. Each set of weights defines the impact of one resolved variable on one demon; resolution of a single variable can affect a number of demons in different ways because independent sets of weights define all demon-variable relationships.

Inference Mechanism

Although ALEX's inference engine does not fit neatly into a rule-based paradigm, POTW EXPERT's mode of reasoning can be described as a combination of backward and forward chaining. One of the defining characteristics of the CPE/CCP methodology is that it investigates all 66 of the Performance Limiting Factors. That is, POTW EXPERT does not employ backward chaining inference to reason from detailed information on plant design and operation back to the one or few factors that are the primary causes of degraded POTW performance. In order to achieve a definitive categorization of the impact of each PLF, POTW EXPERT actively solicits user information on each factor; this process is essentially forward chaining. Process Applications has found this problem-solving methodology to be important for two reasons: (1) systematic investigation of all PLFs often discloses impacts on POTW performance that would not be discerned by simply reasoning backward from symptoms to causes; (2) information on aspects of POTW operation that are functioning well is as important as information on the direct causes of poor performance (e.g., in assuring operators that process capacity is sufficient or that maintenance procedures are adequate to sustain desired plant performance).

After completion of Data Entry and the Major Unit Process Evaluation, POTW EXPERT has accumulated a large body of evidence, stored in variables, relating to the Performance Limiting Factors. POTW EXPERT uses this information to complete a preliminary assessment and classification of the PLFs; this is essentially a forward chaining process, as detailed design and operating information is evaluated to identify each PLF as a candidate for further investigation (the PLF is potentially Class A, B, or C) or to exclude it from further consideration (the PLF receives "No Rating"). After this preliminary assessment is completed, POTW EXPERT follows a forward chaining paradigm to complete its investigation of each unresolved PLF. Investigation continues until the system has enough evidence to make a definitive classification of the PLF.

Each ALEX demon "watches" one or any number of variables. Resolution of variables attached to the demon affects the level of "belief" that the demon is related to poor POTW performance. Analogous to an estimate of probability, belief may take a value between zero and one. Variable resolution affects a demon's belief through a system of weights; each demon holds a listing of the variables it watches, and of the weights attached to each possible resolution of these variables. Weights greater than one increase a demon's belief, while weights less than one decrease the demons's belief. When a variable is resolved, it broadcasts its resolution throughout the ALEX expert system. Each watching demon notes that the variable has been resolved, attaches the appropriate weight, and recalculates its belief.

Backward chaining inferencing in POTW EXPERT proceeds upon completion of Data Entry and the Major Unit Process Evaluation. A large number of variables have been resolved at this point (either by user input or by calculations completed in the Major Unit Process Evaluation), and POTW EXPERT uses these variables to complete as much reasoning about each of the 66 PLFs as it can. For some number of PLFs, this process allows a definitive classification according to their impact on POTW performance, and investigation of these demons terminates. If a demon cannot be definitively classified at this point, however, its belief will be such that ALEX will follow a forward chaining paradigm to solicit additional information (i.e., attempt to resolve additional variables watched by the demon) until a definitive classification can be made.

Example: Aeration Basin Design Performance Limiting Factor. Aeration Basin Design provides a very straightforward example of POTW EXPERT's use of ALEX demons and variables to inference about PLFs. This PLF makes the following assessment:

Does the type, size, shape, or location of the aerator hinder its ability to adequately treat the sewage and provide for stable operation?

This PLF is represented in POTW EXPERT by a context containing two demons that look for faults in the aerator. Both demons watch the same variable, AerationBasinType, output from the Major Unit Process Evaluation; this variable, in turn represents a summary of evidence on adequacy of the type, size (i.e., total oxygen transfer capacity), shape, and location of the aerator to handle loads placed upon it. One demon (AerationBFactor) watches for a Type 2 aerator classification (AerationBasinType=2; aerator marginally capable of handling influent sewage), and one demon (AerationAFactor) watches for a Type 3 aerator classification (AerationBasinType=3; aerator inadequate to handle influent sewage). If AerationBasinType equals 2, then the belief level of the demon AerationBFactor increases, and the belief of AerationAFactor simultaneously decreases; aeration basin design is classified as a Class B Performance Limiting Factor. If AerationBasinType equals 3, however, then the belief level of AerationBFactor decreases and the belief of AerationAFactor increases; aeration basin design is classified as a Class A PLF. If AerationBasinType equals 1 (aerator capacity more than adequate to handle sewage loads), then the belief level of both demons AerationAFactor and AerationBFactor remains low, and the aeration basin design PLF is assigned a classification of "NR", ("No Rating"), and is not cited as a factor potentially limiting plant performance.

Quality Assurance

The Quality Assurance (QA) phase has been an integrated part of the system development process. The QA phase is designed to verify that the system reliably reflects the design specifications and methodology. This analysis has been performed on an ongoing basis by the system developers and a Quality Assurance individual, and through formal, structured reviews by the domain expert. The QA phase is also designed to validate the results and the reasoning of the system. Validation has been performed against test cases designed to exercise the system's critical areas of analysis and reasoning, and against actual case data obtained during two all-day workshops involving a total of forty-three wastewater professionals. Followup with many of these individuals has provided additional input to the validation process.

Conclusion

The POTW EXPERT system demonstrates that a microcomputer-based expert system can effectively represent a complex evaluation methodology, evaluate the capability of a secondary wastewater treatment facility's major unit processes, detect factors which potentially limit performance, and categorize them according to their influence on plant performance. The model is presented in a logical and structured manner to allow wastewater professionals unfamiliar with the CPE process to effectively employ this wastewater treatment methodology.

Acknowledgments

Development of POTW EXPERT has been funded by the United States Environmental Protection Agency under Contract 68-C8-0014, Work Assignment No. 0-01. The contents of this article, however, reflect the views of the authors and do not necessarily reflect those of the U.S. Environmental Protection Agency. Mention of trade names, products, or services is not, and should not be interpreted as conveying official EPA approval, endorsement, or recommendation.

Literature Cited

1. _The 1986 Needs Survey_. U.S. Environmental Protection Agency, Office of Municipal Pollution Control: Washington, DC, 1986.

2. _Handbook: Improving POTW Performance Using the Composite Correction Program Approach_. U.S. Environmental Protection Agency, Center for Environmental Research Information: Cincinnati, OH, October 1984.

RECEIVED March 8, 1990

Chapter 10

The Activated Sludge Advisor Prototype

Joseph Schmuller[1] and Michael R. Morlino[2]

[1]Expert Systems Team, CDM Federal Programs Corporation, 13135 Lee-Jackson Memorial Highway, Fairfax, VA 22033
[2]Camp Dresser and McKee, Inc., Dallas, TX 75231

The Activated Sludge Advisor Prototype (ASAP) is an Expert System designed to assist wastewater treatment operators at the Dallas Water Utilities Central Wastewater Plant. As its name implies, ASAP's focus is the activated sludge process, a biological method that uses microorganisms to speed up decomposition of wastes in wastewater. This process is routinely susceptible to a large number of operational problems. Solving these problems is often difficult for novice operators, particularly if there are no experienced operators around to help. ASAP's knowledge base contains the heuristics and insights of an experienced operator. Written in KnowledgePro (r), ASAP resides on an IBM PC/AT (or compatible); it incorporates a window-based user-interface, graphics, and hypertext. Starting from a set of a half-dozen treatment plant symptoms, ASAP leads the user through sequences of questions resulting in selection of one of 50 diagnoses.

Wastewater Treatment

A wastewater treatment plant is a complex structure with a simple goal -- to insure that dirty water coming in (the influent) will be turned into clean water going out (the effluent). The structure encompasses a number of instruments, processes, and people, all of which are needed because of the wide variety of substances which might contaminate the water.

A treatment facility is usually considered to have three major components -- collection, treatment, and disposal. Wastewater is collected and brought to the facility through a complex network of pipes and pumps; the system which brings water from households, commerce, and industry is typically separate from the system which brings storm runoff water from streets, land, and roofs of buildings.

0097–6156/90/0431–0139$06.00/0

When the wastewater arrives at the treatment plant, it moves through a series of processes which remove the waste from the water and reduce its threat to public health. Treatment at the plant consists of pre-treatment, primary treatment, and secondary treatment. Pre-treatment physically screens out large debris, and removes sand, gravel, and oil. Primary treatment removes settled and floating materials. Secondary treatment involves biological, chemical, and physical processes which remove suspended and dissolved solids; secondary treatment also kills pathogenic organisms.

The Activated Sludge Process

Activated sludge consists of particles produced in wastewater by the growth of organisms in the presence of dissolved oxygen. The term "activated" reflects the high density of bacteria, fungi, and protozoa on the particles. The activated sludge process is a secondary treatment which uses these microorganisms to speed up the decomposition of wastes. When activated sludge is added to wastewater, the microorganisms feed and grow on waste particles in the wastewater, and groups of them come together to form clusters called "floc." They use the wastes for food and as a source of energy for their life processes and for the reproduction of more organisms, which will in turn use more of the waste for food. Also, the activated sludge forms a lacy network, called a "floc mass", that entraps many materials not used as food.

In a treatment plant, the influent flows continuously into an aeration tank, where air is injected to mix the activated sludge with the wastewater and to supply the oxygen needed for the microorganisms to decompose the waste. The mixture of activated sludge and wastewater in the aeration tank is called "mixed liquor". The mixed liquor flows from the aeration tank to a "secondary clarifier", a tank adjacent to the aeration tank, where the activated sludge is settled. Most of the settled sludge is returned to the aeration tank to maintain a high population of organisms to break down the wastes. As more activated sludge is produced than can be used in the process, some of the return sludge is disposed of (wasted).

To keep this process moving smoothly, a wastewater treatment operator has to perform a delicate balancing act across several factors. One of these factors is the food-to-organisms ratio, which is maintained by wasting. Another is the amount of dissolved oxygen (DO) in the aeration tank. If DO is too low, filamentous bacteria could develop, which would keep the sludge floc from settling in the secondary clarifier; if DO is too high, pinpoint floc (tiny clusters of organisms) will develop and not be removed in the secondary clarifier. One final factor is the need to keep flow distributed evenly among multiple treatment units.

In controlling the process, an experienced operator looks for certain signs like the color of the sludge and the amount of turbulence at the surface of the mixed liquor. An experienced operator might also use his/her sense of smell. In addition to these personal observations, an operator must also analyze samples of wastewater taken from different locations throughout the process to confirm his/her judgments. A fair amount of expertise, then, is needed to maintain smooth operation of an activated sludge process.

ASAP

The Activated Sludge Advisor Prototype (ASAP) was designed to help the Dallas Central Wastewater Plant's novice operators control the activated sludge process. Our ultimate goal is to develop an expert system which deals with all aspects of plant functioning. The activated sludge process was chosen as the focus of our initial effort because (a) it is crucial to treatment plant functioning, (b) it entails a considerable amount of expertise, and (c) its circumscribed nature would permit development of an expert system within a reasonable time frame.

ASAP was developed by CDM FPC's Expert Systems team, through a contract held by Camp Dresser & McKee-Dallas with the Dallas Water Utilities. The domain expert, an employee of Camp Dresser & McKee-Dallas, has over 20 years' experience in the wastewater field, and works closely with Dallas Water Utilities. Development proceeded according to well-known expert system development guidelines, facilitated by the wastewater treatment experience of one of the knowledge engineers on the team. We estimate that her experience saved about 2 person-months in the 12 person-month development effort.

Preliminary knowledge was acquired by reading Water Pollution Control Federation publications, and wastewater operator instruction manuals (1), enabling the knowledge engineers to use wastewater terminology when interviewing the domain expert. Before interviewing this expert, the knowledge engineers interviewed a Camp Dresser & McKee engineer who has extensive experience in the design of wastewater treatment plants. Also, prior to knowledge acquisition, the development team toured a treatment facility in Virginia, and the Central Wastewater Plant in Dallas.

Two 3-day knowledge acquisition sessions were held with the domain expert. Three knowledge engineers were present at the first session, two at the second. After the first session, the knowledge engineers' notes were compiled and used as the basis for constructing a "decision diagram" (also called a "knowledge tree") -- a picture which illustrates the sequence of an expert's decisions in given situations, and can be converted into a set of production rules. In the six weeks between sessions, the domain expert was sent the decision diagram for comments and additions, and coding was initiated. At the second session, the decision diagram was used as the basis for interviewing the expert.

The decision diagram grew both in depth and in breadth as a result of the second session. In its final form, the diagram can take one of six paths, each of which has several possible branches. Each path starts with a readily observable symptom in the activated sludge process and ultimately diverges into several diagnoses, so that ASAP can conclude with any of 50 diagnoses. Thus, although ASAP wears the "prototype" label, the depth and breadth of its knowledge base take it beyond the level of a prototype. The specificity of some of the parameter values in its knowledge base confines it to decisions for the Central Wastewater Plant, but it could be customized for other facilities.

The almost-finished system was shown to Central Wastewater Plant operators, so that the knowledge engineers could incorporate target users' comments on the user interface, and their impressions about interacting with the system. The finished system and the final

decision diagram were given to another Camp Dresser & McKee employee
who has extensive wastewater experience. This expert worked with the
system and checked its rules; he was not involved with the development
of ASAP in any other way.

ASAP Implementation

ASAP was implemented in KnowledgeGarden's PC-based expert system shell
called KnowledgePro (r), and it works on a PC/AT or equivalent
machine. We selected this shell because of its flexible knowledge
representation, its interface to graphics packages, its use of
hypertext, and its facilities for easily developing a user-friendly
interface. Our version of KnowledgePro was written in Turbo Pascal,
which can be used as an extension language. Originally, we thought
we might have to do some Pascal programming to implement a record-
keeping capability for ASAP (i.e., to keep track of the user's
observations, user-system interactions, system recommendations, user's
actions taken, and consequences of those actions), but it turned out
that KnowledgePro itself was equal to the task. One drawback of
KnowledgePro (which did not affect this project), is that it does not
contain a built-in facility for implementing certainty factors
(quantitative representations of how sure an expert is of his/her
conclusions); CDM-FPC's Expert Systems Team is currently developing
a way to do this in KnowledgePro for another project.

Knowledge representation in KnowledgePro is built around the
topic, a software structure which has been called the "ultimately
flexible knowledge representation scheme" (2). KnowledgePro uses
topics for storing information, for organizing structure, and as the
basis for hypertext; topics can also be used to set up frame-like
structures which can be used in an object-oriented fashion. Decision
points on the ASAP decision diagram were easily and intuitively
represented as KnowledgePro topics, and options at each decision point
were easily represented within the topics. Production rules (the
expert's "heuristics") were also represented within topics, providing
organization to the knowledge base. Unlike other expert system
development tools (such as OPS5), a production rule's location (with
respect to topics) is of extreme importance in KnowledgePro.

ASAP incorporates hypertext, a technique which connects pieces
of text in a non-linear fashion. Hypertext allows the developer to
link a term or a phrase to its definition or to other clarifying
information, so that the additional information is hidden from view
when the term is on the screen; the user, at his/her option, can bring
the additional information onto the screen. A term can be linked to
a descriptive graphic, as well as to text. The additional information
can be similarly linked to more information, and so on. ASAP
incorporates all of these techniques, resulting in a user interface
which each user, in effect, tailors to his/her own capabilities during
each session. That is, each user has a potentially different
background and level of experience, and will know the definitions of
some terms and not others. Hypertext provides an opportunity for a
user to look up a definition without breaking away from the system.

The general development of the user interface proceeded along
with the development of the system. Ordinarily, with any of a number
of other expert system shells, we would have first developed the
knowledge base, tested it, and then developed the user interface.

KnowledgePro, however, enables the developer to construct a window-based interface very easily and quickly. We believe that developing the user interface in this way (i.e., while implementing the system as a whole) resulted in a time saving of about 10%. We also feel that it represents a more natural way to develop a system.

A session with ASAP begins with the user entering identification information and initial comments about observed conditions. The session proceeds with ASAP questioning the user, and the user answering by choosing one of several options, except when ASAP directs the user to enter quantitative information. As shown in Figure 1, each question appears in a window, and answering options appear on a menu within the window. Terms linked to their definitions via hypertext are highlighted, so that the user always knows which terms have further information available; function keys allow the user to navigate among these terms and display their definitions. Some of the terms are linked to text, some to graphics. As Figure 2 illustrates, a text-based definition or clarification, when brought to the screen, appears in a window which is smaller than a question-window, and partially overlaps with the window which links to it. Graphic hypertext dominates most of the screen. ASAP's conclusions also appear in windows. A session concludes with the user entering comments, actions taken, and consequences of those actions. The entire set of questions, responses, and comments in a session is then written to a file.

Summing Up

At present, wastewater operator tests of ASAP are pending. Management staff at Dallas Water Utilities have run through sessions with ASAP, and their reaction has been very positive.

We envision several future directions for our work with ASAP. First, we think that ASAP could be used as a training tool as well as a troubleshooting tool. In the short term, this will occur as operators become increasingly comfortable with ASAP, and enter hypothetical wastewater plant situations, as well as situations which they have observed. The operators' noting and remembering ASAP's responses will effect a transfer of expertise from ASAP's knowledge base (and, ultimately, from our domain expert) to the operators. In the long term, ASAP could serve as one component of an Intelligent Computer-Assisted Instruction (ICAI) system. ICAI systems combine the knowledge of a domain expert and the knowledge of an experienced teacher, resulting in a computerized tutor in a particular field. A user interacts with an ICAI system in much the same way that he/she would interact with a human tutor -- by asking questions, solving problems posed by the tutor (which are geared to the user's mastery of the field and learning rate), and having his/her problem solutions evaluated by the tutor.

We also think that a logical next step for ASAP is to expand the knowledge base to deal with trends in the mass of data which is continuously being gathered at the Central Wastewater Plant. As KnowledgePro has database interface facilities, we feel that this capability could be implemented very easily.

Finally, we believe that ASAP will serve as the cornerstone for reaching our aforementioned goal of a system which helps operators to control and troubleshoot all phases of wastewater plant operation,

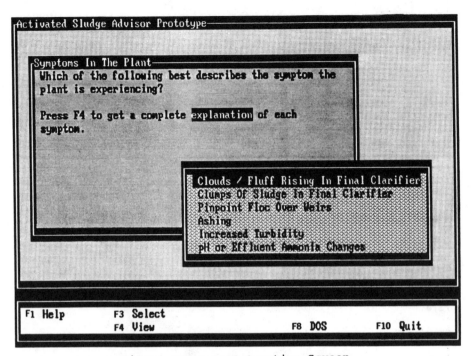

Figure 1. An ASAP Question-Screen

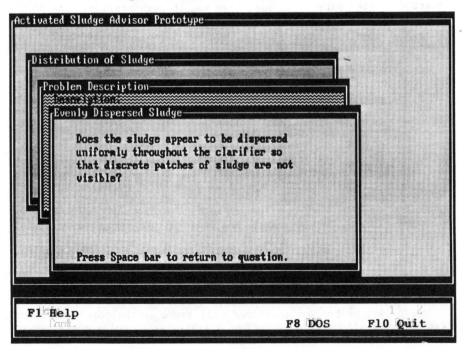

Figure 2. ASAP Hypertext

including pre-treatment, primary treatment, secondary treatment, and disposal of removed wastes. Such a system might someday be interfaced to devices for automatic process control, thus automating many of the operations of a wastewater plant.

Literature Cited

1. Operation of Wastewater Treatment Plants: A Field Study Training Program, Volumes 1 and 2, U. S. Environmental Protection Agency, Office of Water Program Operations, 1988.

2. Shafer, D. PC AI 1988, 2(2), 37-39.

RECEIVED April 13, 1990

Chapter 11

Multidomain Expert Systems for Hazardous Waste Site Investigations

H. Y. Fang[1], G. M. Mikroudis[2], and S. Pamukcu[1]

[1]Geotechnical Engineering Division, Department of Civil Engineering, Lehigh University, Bethlehem, PA 18015
[2]Roy F. Weston, Inc., 955 L'Enfant Plaza, S.W., Sixth Floor, Washington, DC 20024

A new tool for investigation of hazardous waste sites and other related geo-environmental problems is introduced. This tool is a multi-domain expert system which is an extension of an existing knowledge-based expert system, namely GEOTOX. The new system has additional capabilities such as management of larger amount and diversified data by use of modular expert systems (MES). Computer integration and user interaction is enhanced through colorful displays and/or tabular results. These features lead to a greater degree of unification in planning, analysis and management process across many disciplines related to hazardous waste site investigations.

In recent years, due to population growth, a progressive living standard and industrial progress, air, water and land have become polluted. Open dumps and chemical and industrial wastes cause problems such as those listed in Table 1. Improper disposal and management of hazardous wastes and toxic chemicals in numerous unidentified locations is one of the most pressing environmental problems at the present time. In 1978, the U.S. Environmental Protection Agency (EPA) estimated that approximately 60 million metric tons of hazardous waste are generated annually in the U.S. at more than 750,000 sites. Figure 1 shows various types of wastes generated in some developed countries and Table 2 shows the typical hazardous substances in a selection of industrial wastes.

0097–6156/90/0431–0146$06.00/0
© 1990 American Chemical Society

TABLE 1   SOME CAUSES OF GROUND SOIL POLLUTION IN U.S.A.

Southwest	South Central	Northeast	Northwest
Acid Rain	Acid rain	Acid Rain	Abandoned Oil Wells
Animal Wastes	Animal Wastes	Acid Mine Drainage	Acid Drainage & Mine Tailing
Disposal of Oil Field Brines	Disposal Well	Buried Pipelines & Storage Tanks	Acid Rain
Evapotranspiration from Vegetation	Evapotranspiration from Vegetation	Hazardous Chemical Wastes	Brine Injection
Hazardous Chemical Wastes	Hazardous Chemical Wastes	Highway Deicing Salts	Disposal Well
Injection Wells for Waste Disposal	Irrigation Return Flow	Landfills	Dry Land Farming
Irrigation Return Flow	Landfills	Mine Fire	Highway Deicing Salts
Leaching	Nuclear Wastes	Nuclear Wastes	Hazardous Chemical Wastes
Nuclear Wastes	Oil Field Brines	Petroleum Exploration & Development	Irrigation Return Flow
Saltwater Intrusion	Solid Wastes	Radon (gas)	Landfills
Solid Wastes	Waste Lagoons	River Infiltration	Mine Fire
Spills Hazardous Material		Saltwater Intrusion	Nuclear Wastes
Water from Fault Zones & Volcanic Origin		Septic Tanks	Septic Tanks
		Surface Impoundments	Swage Treatment Plant Discharges
			Surface Impoundments

TABLE 2. Typical Hazardous Substance In Industrial Waste Streams

Industry	Arsenic	Heavy Metals	Chlorinated Hydrocarbons	Mercury	Cyanides	Selenium	Misc. Organics
Chemical	---	X	X	X	---	---	X
Electrical & electronic	---	X	X	X	X	X	---
Electroplating & metal industry	X	X	---	X	X	---	X
Leather	---	X	---	---	---	---	X
Mining & metallurgy	X	X	---	X	X	X	---
Paint & dye	---	X	---	X	X	X	X
Pesticide	X	---	X	X	X	---	X
Pharmaceutical	X	---	---	X	---	---	X
Pulp & paper	---	---	---	X	---	---	X
Textile	---	X	---	---	---	---	X
MSW	X	X	X	X	X	X	X

Misc. organics include various phenols, benzenes, etc.

SOURCE: **Reprinted with permission from ref. 13. Copyright 1989 Envo Publishing Company, Inc.**

PERCENT OF TOTAL WASTES GENERATED

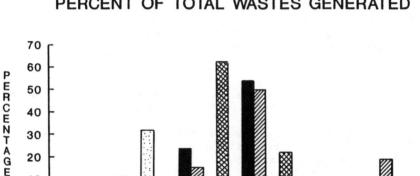

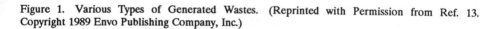

Figure 1. Various Types of Generated Wastes. (Reprinted with Permission from Ref. 13. Copyright 1989 Envo Publishing Company, Inc.)

The objectives of this paper are (1) to introduce a newly developed multi-domain knowledge-based expert system as an aid for hazardous waste site investigations, and (2) to develop a better knowledge-based database and data management system.

Basic Information Required for Hazardous Site Investigations

The description and quantification of actual and potential hazards associated with a waste disposal site incorporate elements of site evaluation, chemical fate and transport evaluation, basic toxicology, exposure, and risk assessment(1). Many factors must be considered before a site characterization is possible. Assessment of the hazard level is a complex engineering problem which requires interdisciplinary knowledge. Understanding of the interaction between toxic wastes, water and soil/rock is based on concepts of environmental geotechnology, geology, hydrology, climatology, chemistry and toxicology. Basic information required for hazardous site investigations is discussed as follows (2),(3):

Sources. The major sources of soil/water pollution are listed in Table 1. Regardless of the sources, there are three basic mechanisms by which the ground soil/water can be contaminated:

(1) Contamination may occur from rainfall such as acid rain, or rain falling into a sanitary landfill, or oil or chemical wastes spilled onto the soil/water systems (4).

(2) Leakage of pollutants from disposal wells or constructed waste disposal facilities such as landfills, septic tanks, laterals and lagoons.

(3) Hydraulic, chemical, or physico-chemical alterations which allow pollution substances to move within or between soil layers. In this category, the phenomena cover chemical, physicochemical, and microbiological aspects (5).

Receptors. To identify potential receptors and the contaminant levels to which they are exposed, the following points where exposure to contaminants may occur are commonly evaluated: contact, wastes, and fish.

Pathways. The pathways of exposure in terms of populations and/or environments around the site that may be affected via that pathway. Consideration is given to the following factors: profile of soil layers, hydraulic conductivity, discharge, use, and habitats.

Once the pathways, receptors and sources of contamination have been identified and evaluated, various methodologies can be applied to give a quantitative or qualitative measure of the potential treats to human health, welfare or the ecosystems. In all cases, a series of investigations and data collection activities is required before one can proceed to the site evaluation.

A complete assessment of a hazardous waste site requires consideration of not only th technical aspects but also non-technical aspects such as political/social/economical which are also important. In many cases, these non-technical concerns are even more important than the technical ones. Let's examine the whole assessment system which covers three major phases: the transverse, diagonal and longitudinal interactions, as illustrated in Figure 2. On the surface, these three phases are not directly interrelated, however, without considering each phase, the assessment of site investigation cannot be effectively undertaken. Without political/social understanding, the technical part is only a small part of an overall system. In order to provide for the efficient, economical results, it is necessary to have a well-planned system. Some of the important factors affecting a hazardous site investigation are discussed in the following text.

Transverse Interaction. This phase covers most of the nontechnical factors such as: political (legislation, zoning, etc.), public opinion or news media, and social tradition. These non-technical factors are the major decision-making factors, for example, for determining whether or not this site investigation will be funded by an EPA superfund. These factors, in some cases, may be more important than the technical factors.

Diagonal Interaction. This phase includes mainly the economic factors which include total investment, interest charges, time constraints, availability of labor, and construction equipment and annual maintenance cost.

Longitudinal Interaction. This phase mainly covers the technical part and will be further discussed in the following sections.

Nature of Knowledge-Based Expert Systems

Knowledge-Based Systems (KBS) are computer programs that contain expert knowledge about a specific domain and are able to apply this knowledge to make useful inferences and provide expert-level advice to the user of the system. In addition, expert systems are capable, on demand, to 'justify' their own line of reasoning in a manner directly intelligible to the inquirer.
    The knowledge includes facts and rules - both of which are interchangeable. Facts represent declarative knowledge and provide an actual database. Rules represent procedural knowledge, as well as rules of judgment and plausible reasoning. Inference includes the techniques used to generate new information from old and is applied according to a control strategy, e.g., choosing which rules to apply, and trying different alternatives. The advantages and limitations of KBS and comparisons between conventional programming and KBS are discussed by Fang et al., (3).

The Effectiveness of KBS. The effectiveness of a KBS depends on the system and shell used. If the project is a straight-forward type, you do not need a KBS. If the project is too complicated, an expert system cannot provide effective results. At present, most available systems are the single-domain type (Figures 3 and 4). A single-domain system consists of the following basic parts: the human-computer interaction as shown in Figure 3, and the main portion of the system known as the shell (Figure 3(B) which, in general, consists of artificial intelligence (AI) technology, graphics and risk analysis.

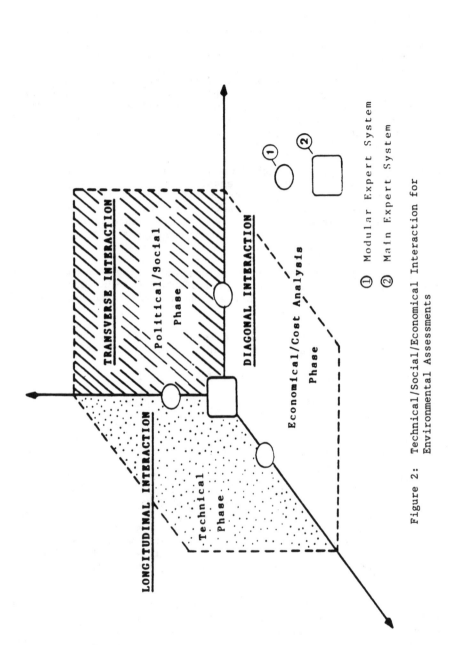

Figure 2:   Technical/Social/Economical Interaction for
            Environmental Assessments

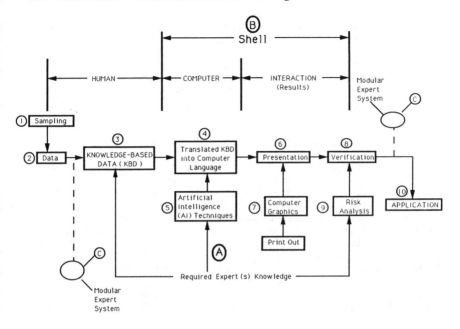

Figure 3: Human-Computer Interaction in Knowledge-
Based Expert Systems

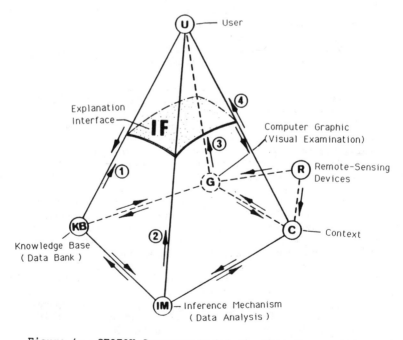

Figure 4: GEOTOX Computer Model for Data Management,
Analysis and Presentation

Each shell is designed for a particular application; some shells accomplish more functions than others (6). There a number of commercially available shells on the market. GEOTOX, developed by the Envirotronics Corporation, is one such shell and is useful for engineering analysis, design, construction and management (3) (7). GEOTOX originally was designed for a main-frame computer system, but since has also been up-dated and converted into a personal computer (PC) system (8).

Structure of the GEOTOX Shell

The theoretical background of GEOTOX shell has been discussed by Fang and Mikroudis (3) in a separate paper. In this section, an outline of the basic features of the GEOTOX shell together with recent improvements on the shell and other related features are presented.

GEOTOX is a knowledge-based expert system primarily designed for hazardous waste site evaluations. More specifically, it is intended to assist in preliminary investigations, although its knowledge base can be expanded to accommodate detailed investigations and field work. It can be used for multiple site comparison and ranking. Besides evaluation of existing sites, GEOTOX can be applied to assess potential sites and to assist in the site selection process for new facilities. The various processes supported by GEOTOX are summarized as follows:

(a) Interpretation: Assessment of existing hazardous waste sites. Evaluation of potential waste disposal sites.

(b) Classification: Ranking of existing sites. Screening of potential sites.

(c) Diagnosis: Contamination problems at hazardous waste sites. Selection of remedial alternatives.

General design requirements for 'expert' systems have been identified by researches in Artificial Intelligence (AI) (9). More recently, the human behavior in engineering or psycho-engineering concepts have been introduced into the 'expert opinion' (8).

Figure 4 shows human-computer interaction using the GEOTOX model. In examing Figure 4, a user (U) communicates with the subsystem of the database (KB), graphics (G) and analysis design (IM) through a human-machine interaction (IF) which is shown by the shaded curved surface. This interface can support 'customized' communications that allow a user to selectively choose from a wide variety of options throughout these subsystems. The linkages for data transfer among these systems are indicated by the lines with arrows. At any stage, the user can trace backwards and forwards to see what has been done.

The main portion of GEOTOX covers the knowledge base, associate network, production rules, frames, and inference rules (10). A simplified view of the knowledge representation scheme is shown in Figure 5. The hybrid system of an associative net and production rules is implemented using logic programming via the computer language PROLOG. The associative network provides the underlying structure to the overall knowledge representation scheme. This network defines all the associations between data and site parameters as conceived by the domain expert(s). The production rules are used at conjuctive nodes (Figure 5) of the network to determine what value should be inherited by the network according to the existing conditions at the node.

Frames (Figure 5) are a collection of associative net nodes and slots. They are used to represent the final conclusions of the expert and to describe different situations or possible scenarios of contamination at the site. The application of the inference rules in GEOTOX is shown in Figure 6. GEOTOX uses the hazard value 'h' and confidence level 'c' as provided by the production rules and propagates them to all the associated nodes following the links in the associative network. Figure 6 (a) and (b) show the 'h'-'c' pair at typical nodes 'i' and 'j' of the GEOTOX associative network. Figure 6 (c) illustrates the effect of applying the inference rule for a given 'h'-'c' pair. It shows that the updated 'h' in Figure 6 (c) is closer to the value provided with more confidence. For detailed theoretical background on GEOTOX, see reference (3). The main features and advantages of the GEOTOX are summarized in Figure 7 and discussed by Fang et al., (3).

Multi-Domain Knowledge-Based Expert System

If a project consists of multi-characteristics such as large structural systems, i.e., tall buildings, offshore structures, or complex projects such as a large construction operation or hazardous waste control facility, a single-domain system may not be used effectively. Therefore, a new multi-domain system (MDS) has been developed (8) as shown in Figure 8.

MDS is an extension of the GEOTOX model, containing additional features for addressing widened applications. The central components of the model are the user-system and system-system interface which provide the links among the diverse software subsystems and establish a seamless transition from application to application. This interface provides facilities for process-to-process communication in addition to the standard facilities for communicating and reporting the processes of the system and the results to the user (as in the GEOTOX model). This interface establishes an open architecture where more satellite subsystems can tie-in and communicate with all the other programs with direct links using the interface facilities. These facilities will include customized menus and programs as well as generic routines available, as a library, that can be included in the new applications.

The main function of the MDS model is to establish an environment that encourages compatible applications. One person's main application program is another's subprocess or subroutine via direct links of the MDS model. In addition, this model will provide a rich toolbox for manipulating data, creating knowledge bases and combining different problem-solving components together. Driven by the MDS interface, new KBES applications will accept input form users, real-time sources, sensors, and data bases; will make all inferences or retrieve all available information; and will request additional information when needed. Because of the complexity of such applications, the interface should provide facilities for generating the code, for testing the consistency of new KBD with old, and for ensuring the concurrency of databases and knowledge-bases.

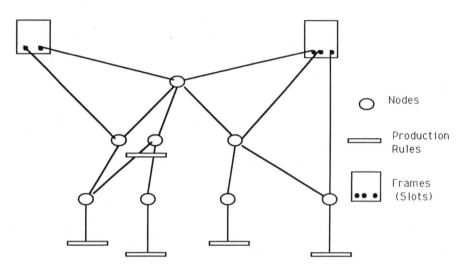

Figure 5:   Simplified View of the Knowledge
Representation Scheme in GEOTOX

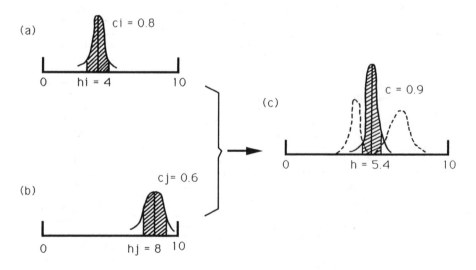

Figure 6:   Application of the Inference Rule in GEOTOX

FEATURES   **BENEFITS**

KB : KNOWLEDGE BASE

O **PRODUCTION RULES** (for expert   Modularity, simple semantics,
derived heuristics)   easier knowledge acquisition

O **SEMANTIC NETWORK** (defines the   Cause-effect relationships,
problem solving strategy, and   classification properties
parameter interactions)   easily described

O **FRAMES** (for conclusions and   Various types of evaluations
recommendations)   and possible situations defined

IM : INFERENCE MECHANISM

O **CONFIDENCE FACTORS**   Expresses confidence in data

O **COMBINATION OF FORWARD**   Handles both interpretive
AND BACKWARD CHAINING   and diagnostic problems

IF : INTERFACE

O **HOW / WHY**   Examine the line of reasoning

O **SUMMARIZE / CONCLUDE**   Review the state of knowledge

O **VOLUNTEER**   Flexible data entry,

O **CHANGE**   Data update "what if " questions

O **REVISE**   Modify the knowledge base

User specified options include :

G : COMPUTER GRAPHICS   Visualization of conditions
DB: DATA BASES   Access to data bases
A : ANALYSIS PROCEDURES   Use of analytical models
R : REMOTE SENSING   Ability to incorporate
                                              remote-sensing devices

Figure 7. Features and Advantages of the GEOTEX Shell.

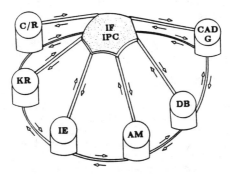

Figure 8: Fang-Mikroudis Model of Multi-Domain
Knowledge-Based Expert System

Key: IF—Explanation Interface,   IPC—Inter-Process Communication,
C/R—Communication/Real Time, KR—Knowledge Representation, IE—Interface Engines,
AM—Algorithmic Model, DB—Data Base Manager, CAD—Computer Aided Design,
G—Computer Graphics.

Thus, a real-world system is needed which is based on such an architecture that will provide all the KBES tools, databases, analytical models, and interfaces to the outside world for building an MDS application. As long as a general protocol for transferring information among applications is established, the whole MDS model does not need to be present for developing practical applications. A partial implementation of subsystems of the model can be combined when needed with other compatible applications at a later stage. In this way, the real-world knowledge that is contained in existing databases and analytical models can be made accessible and can be manipulated by MDS application programs. The expertise of such real-world programs relies on the communication with other machines and applications that provide the data and knowledge necessary to solve the real-world problems.

The main features of the MDS includes:

IF/IPC   - a generalized user-system and system system interface (IF) with built-in interprocess communication channels.

KR   - Several knowledge representation languages and techniques.

IE   - Several inference engines that can be combined with the various KR techniques.

DB   - A database manager that provides access to various databases.

AM   - An analytical/algorithmic model library that allows execution of different algorithmic programs.

C/R   - Communication/real-time access of other computer systems (e.g., via a network), or instruments, sensors, and real-time devices.

G/CAD   - Graphics and computer-aided design packages.

Modular Expert System (MES)

The Modular Expert System (MES) is an integral part of multi-domain expert systems. In multi-characteristic projects such as hazardous waste site investigations, requiring interdisciplinary knowledge, it is difficult to put all the necessary knowledge-based data into one system when using a personal computer. Therefore, a Modular Expert System (MES) is needed to integrate the separate systems (10).

The structure of the MES is similar to the GEOTOX model shown in Figures 3 and 4. The major difference between these two systems is the knowledge-based data bank (see Figure 3, steps 2 to 3 and Figure 4).

The MES is just like a list of menu options. A menu selection is connected with the main multi-domain expert system as needed and disconnected when not needed. In Figure 2, the interactions among technical and non-technical aspects require additional expert systems to incorporate the unrelated systems into a related unit. Such a temporary expert system unit is referred to as Modular Expert System (MES). When organizing a large amount of experimental data into a knowledge-based data system, expert opinion and an expert system is also required. In such cases, a Modular Expert System is needed to integrate the data. Similarly, when the GEOTOX model is used for various applications, each application will be different. Therefore, modifications are needed for each case; the MES can integrate these modifications into a reliable output.

The major function of MES is to aid in solving defined problems under specified conditions. It is a temporary system as far as the main expert system is concerned. In other words, the MES unit is used to custom fit individual or particular projects as needed. Several basic MES units have been developed and are discussed in a separate report (7). A flow chart illustrates how GEOTOX or MES operates during a typical consultation session (Figure 9).

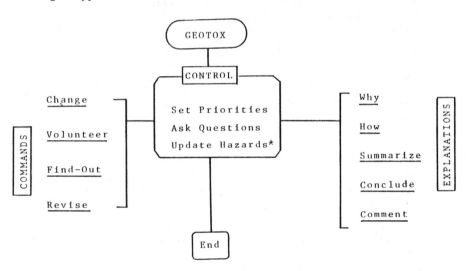

* Other problems as shown in
Figs. 2 and 3.

Figure 9: A Flow Chart Illustrating How GEOTOX or MES
Operates During A Typical Consultation
Session

Summary and Conclusions

1. Assessment of site investigations for hazardous wastes requires interdisciplinary knowledge. The GEOTOX knowledge-based expert system is introduced to manage complex interactions among many variables.

2. In many cases, it is difficult to put all the necessary knowledge-based data into one-system when using a personal computer

(PC); therefore, it is necessary to have Modular Expert Systems (MES) to integrate these into the main system. The structure of the MES is similar to the GEOTOX model but with different knowledge-based databases.

3. The Multi-Domain Expert System is an extension of the GEOTOX model which covers more features and widens the applications of the GEOTOX model. In addition to hazardous site investigation, this model has been used successfully to address various geo-environmental and engineering problems such as radon assessment, landslide control, and large structural systems.

4. Additional functions of a multi-domain system include a database for storage of large amounts of data, classification of data, correlation studies such as that for identifying degree of hazard relating to geotechnical/hydrological data.

5. All information produced with these expert systems includes colorful pictorial displays and/or tabular results at any given stage of interaction. Also, the user can trace back to see what has been done, or may interactively alter technical and/or financial criteria and constraints.

6. The advantage of the computer integrated systems is that they can lead to a greater degree of unification in the planning, analysis, design and management processes across many disciplines.

<u>Literature Cited</u>

1. Berger, I. S. (1984), Determination of Risk for Uncontrolled Hazardous Waste Sites, <u>Proc. National Conf. on Management of Uncontrolled and Hazardous Waste Sites</u>, U.S. EPA, pp. 23-26.

2. Desmarais, A. M. C. and Exner, P. J. (1984), The Importance of the Endangerment Assessment in Superfund Feasibility Studies, <u>Proc. National Conf. on Management of Uncontrolled and Hazardous Waste Sites</u>, U.S. EPA, pp. 226-229.

3. Fang, H. Y., Mikroudis, G. K., and Pamukcu, S. (1987), A Unified Approach to the Assessment of Waste Disposal Sites, <u>Computers and Geotechnics</u>, G. N. Pande, ed. Elsevier Applied Science Publishers Ltd., England, Vol. 3, pp. 129-156.

4. Pamukcu, S., Mikroudis, G. K. and Fang, H. Y. (1987), "GEOTOX" A New Knowledge/Data Base Management System for Controlling Wastes, <u>International Symposium on Environmental Management</u>, General Directorate of Environment-Pollution Control Research Group, Bogazici University, Istanbul, Turkey, pp. 1075-1094.

5. Fang, H. Y. (1989), Expert Systems for Assessment of Radon Gas, <u>ASCE Environmental Engineering National Conference</u>, J. F. Malina, ed. ASCE, New York, pp. 97-104.

6. Feigenbaum, E. A. (1977), The Art of Artificial Intelligence: Themes and Case Studies of Knowledge Engineering, IJCAI 5, pp. 1014-1029.

7. Mikroudis, G. K. and Fang, H. Y. (1987), Classification of Waste Disposal Sites Using GEOTOX, ASCE Geotechnical Practice for Waste Disposal '87, R.D. Woods, ed. ASCE Geotechnical Special Publication No. 13, New York, pp. 105-120.

8. Fang, H. Y. and Mikroudis, G. K. (1987), Multi-Domains and Multi-Experts in Knowledge-Based Expert Systems, Proc. 1st. International Symposium on Environmental Geotechnology, Envo Publishing Co., Bethlehem, PA, Vol. 2, pp. 355-361.

9. Buchanan, B. C. and Shortliffe, E. H. (1984), Rule-Based Expert Systems, The MYCIN Experiments of the Stanford Heuristic Programming Project, Addison-Wesley Publishers.

10. Barnett, V. (1973), Comparative Statistical Inference, John Wiley & Sons, N.Y.

11. Fang, H. Y. and Mikroudis, G. K. (1989), Modular Expert System: An Expert System Between Expert Systems. Report prepared for Envirotronics Corporation, International.

12. Fang, H. Y. (1987), Soil-Pollutant Interaction Effects on the oil Behavior and the Stability of Foundation Structures, in Environmental Geotechnics and Problematic Soils and Rocks, A. S. Balasubramaniam, et al. eds. A. A. Balkema Publishers, pp. 155-163.

13. Yong, R. N. (1989), Waste Generation and Disposal, Proc. 2nd International Symposium on Environmental Geotechnology, Vol. 1, Envo Publishing Co., Bethlehem, PA, pp. 1-24.

RECEIVED April 27, 1990

Chapter 12

The Cost of Remedial Action Model

Expert System Applications

Marie T. Chenu and Jacqueline A. Crenca

CH2M Hill, P.O. Box 4400, Reston, VA 22090

The Cost of Remedial Action (CORA) model was developed
for the United States Environmental Protection Agency
(EPA). The model is used in developing and costing
remedial actions for Superfund sites before or during
the remedial investigation of the cleanup.

The   CORA   model   includes   two   independent,
microcomputer-based subsystems.   One subsystem is a
knowledge-based   consultation   program   that   develops
remediation recommendations. The second subsystem is a
database management system that develops site-specific
cost   estimates   for   the   technologies   required   to
implement the expert system's recommendations. Use of
the model has made possible a considerable time saving
over manual scoping and costing, and has also led to
consistent procedures for remedy selection across EPA
regions.

This paper discusses the expert system and its development, testing, validation, and application.

The Cost of Remedial Action (CORA) model was developed for the
United States Environmental Protection Agency (EPA). The model is
used in developing and costing remedial actions for Superfund sites
before or during the remedial investigation of the cleanup.

System Capabilities

The CORA model was designed to run on an IBM-compatible microcomputer. It requires 587K of free RAM (beyond the RAM required by the
operating system) and a minimum of 5 megabytes of free disk space.

0097–6156/90/0431–0162$06.00/0

The model runs in either a color or a monochrome mode, depending on the monitor.

The CORA model includes two independent, microcomputer-based subsystems. One subsystem is a knowledge-based consultation program developed with the Level 5 expert system shell, dBASE III+, and Nantucket Clipper. This subsystem comprises four knowledge bases that communicate with each other and update facts during execution. The second subsystem is a database management system written in dBase III+ and Nantucket Clipper that develops site-specific cost estimates for the technologies required to implement the expert system's recommendations. The estimate is that use of the model has made possible a 5-fold to 15-fold time saving over manual scoping and costing. Use of the model has also led to consistent procedures for remedy selection across EPA regions.

The expert system analyzes a site by focusing on separate user-defined contaminated areas of the site. For each contaminated area, the user may designate up to 13 waste types, from buried drums to contaminated saturated soils. After an interactive dialog with the user, the expert system produces a summary report that lists a range of potentially implementable and applicable remedial action (RA) technologies for each waste type. The technologies, which include both treatment and containment technologies and range from using vapor-phase carbon to asphalt caps, can be combined by the user to form one or more RA alternatives.

For each question asked by the expert system, the user is presented with a menu, or list, of possible answers (mainly in "True/False" format) from which to choose, or the user may be asked to enter textual information, using the keyboard. The user interface also includes extensive on-line help for first-time or infrequent users who may not be familiar with expert system terminology or with the user interface of the Level 5 expert system shell. This "help" feature can be invoked at any point during a consultation by pressing a function key.

After RA scenarios are determined for the site, the cost system develops estimates that have a target accuracy range of +50 to −30 percent of actual costs. The expert system comprises 40 technologies covering containment, removal, treatment, and disposal options. In addition to containing cost algorithms for each of the technologies that may be recommended by the expert system, the cost system develops cost estimates for site preparation, site administration, health and safety, and contingencies and allowances.

The system is not designed to incorporate all of the many technologies that would be necessary to address every type of site; instead, the goal is to address the majority of sites. "Outliers" include mining sites and sites containing radioactive waste. Some emerging technologies, such as in situ vitrification or ultraviolet ozonation, were not included in the model because of their scope and uncertainties. The CORA framework, however, allows for expansions, and other technologies will be considered for addition during annual updates of the model.

Development of the Expert System

Overview. The CORA expert system has gone through two major phases in its development. The first-phase deliverable was a prototype system. The knowledge bases and explanation facility were written using the Production Rule Language (PRL) of Insight 2+. Additional user interface and reporting capabilities were developed in dBASE III+ and Nantucket Clipper.

The prototype contained 550 rules in six knowledge bases dealing with contamination in groundwater, lagoons and ponds, and soil. The prototype was designed to be used by a CH2M HILL representative trained in using the model and knowledgeable about Superfund sites, working with an EPA regional project manager (RPM). On the basis of an interactive consultation, the system could recommend up to 28 remediation technologies.

In the second phase of the project, EPA wanted to distribute the model to the EPA regions for independent use, and CORA was refined and upgraded to facilitate that use. The expert system was revised to include four knowledge bases containing 671 rules and up to 40 technologies for recommendation. Insight 2+, whose name was changed to Level 5; dBASE III+; and the Nantucket Clipper compiler were used.

The second version of the expert system was completed in approximately 8 months and was released in June 1988. EPA and the U.S. Navy have used it to select and cost RAs under their fiscal year 1989, 1990, and 1991 budgets for remediating hazardous waste sites.

A third version of CORA is scheduled for release in early 1990. It will contain updates of both the expert system and the cost system.

System Goals. From the beginning, this project was shaped by several important constraints. First, the system's recommendations had to be based on current EPA policies and technological considerations, so the system had to be designed to allow changes as both policies and technologies evolve. This constraint significantly influenced our decision to select a knowledge-representation scheme based on production rules, where each rule is constructed in a simple "If . . . then . . ." format, the premise being a Boolean expression and the action containing one or more conclusions. Each rule is modular and independent of the others. From a development standpoint, such modular coding leads to a relatively uncomplicated control structure (1) that has the benefit of allowing new rules to be added easily.

The second constraint was that the first prototype needed to be ready for field testing in less than 6 months. This constraint led us to choose an expert system shell, because a shell would have the inference engine and user interface built in, permitting significant time saving in comparison to the time required for implementing an expert system with a language such as Lisp or Prolog.

Our third constraint was the need to design a system that would run on a microcomputer with only the basic 640K of RAM. This need stems from two facts: Most of EPA's microcomputers had that hardware configuration, and EPA wanted the maximum number of users to have access to CORA without incurring the high cost of option-laden microcomputer hardware or the connect-time costs and possible lack of access of a mainframe system. Finally, the fourth constraint was that the chosen software must not levy licensing fees on production copies.

To meet these constraints, we chose for our development tool Insight 2+, a rule-based and microcomputer-based expert system shell that met most of the criteria of a good tool, as described by Waterman (2)--that is, it has good user interface, rule-tracing, and debugging capabilities. In addition, Insight 2+ has no requirements for licensing fees.

<u>Knowledge Acquisition and Implementation</u>. The domain knowledge in CORA was obtained from the experiential knowledge of CH2M HILL's environmental engineering experts, information on current EPA policies derived from interviews with senior EPA managers, and from EPA reports (3), the Superfund Amendments and Reauthorization Act (SARA), and the Hazardous and Solid Waste Amendments (HSWA).

The approach of acquiring and implementing knowledge that was used in the first phase of development was curtailed by time constraints. In developing the necessary rules, knowledge engineers relied primarily on decision trees drawn after interviewing the domain experts. Most of the rules were linked in some way. This approach resulted in a very fast system that functioned much like a conventional program and exerted strong control over the inference engine. Predictably, the system was apt to fail whenever the user tried to branch to a path not defined by the rules.

In the second phase, a less deterministic approach to knowledge implementation was chosen. A set of preliminary rules in the form of "If . . . then . . ." statements was developed, and the set was grouped into categories such as landfills, above-ground contamination, and removal options and issues. The knowledge engineers rewrote the rules using PRL syntax, and these new rules formed the knowledge bases. Working closely with the domain experts, the knowledge engineers then refined the existing rules and added new rules by extensively testing and running the knowledge bases. The advantages of using a rule-based system simplified the communication between knowledge engineers and domain experts. Because rules are a straightforward way of expressing knowledge, understanding them did not require knowledge of a programming language. By following the inference chain and looking at each rule, the experts could easily understand the system's line of reasoning and could suggest modifications during testing.

A major difficulty encountered during implementation was the way in which recommendations had to be presented to the users. For the system to be as user friendly as possible, system recommendations had to be grouped into categories such as treatment, containment, and discharge, rather than in the order in which the recommendations were deduced. Another reporting criterion was that multiple recommendations for RA technology that fall under one category heading be shown in the "Either . . . or . . ." format. Shown below is an actual expert system recommendation under the heading "In Situ Soil Treatments":

Either
o   Soil-vapor extraction for VOCs
       Either
       o   Flare for VOCs
       Or
       o   Vapor-phase carbon
Or
o   Deal with the principal threat (i.e., buried tanks or drums)

Level 5 had no facility to handle these reporting requirements adequately, so a significant part of the implementation period was spent working on the strategy for representing such "nested" levels of recommendations. Additional rules had to be added to write the recommendations to disk, and dBASE III+ programs were written to sort and present the recommendations to the users. See Figure 1 for an overview of the expert system architecture.

The CORA expert system was developed by a team of three knowledge engineers, two of whom were programmers and one of whom was a domain expert. The latter was responsible for interviewing other domain experts. Numerous other domain experts also served as reviewers. An example of output from an expert system run is shown in Figure 2.

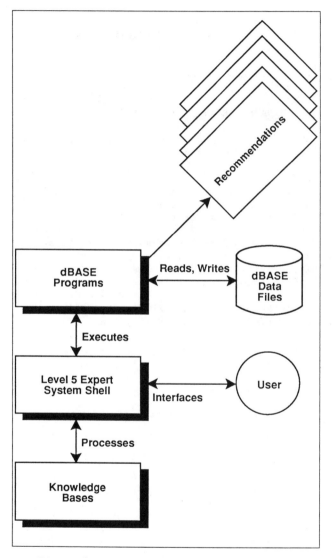

Figure 1.   Overview of CORA Expert System

CORA EXPERT SYSTEM

RUN: TEST RUN FOR ACS
RUN BY: M. CHENU
SITE: TEST SITE
CONTAMINATED AREA: CONTAMINATED AREA 1

WASTE TYPE: HOT SPOTS (UNSATURATED MTL AROUND
LEAKY TANKS OR DRUMS)

INPUT
Response type: Treatments
Soil description: Medium sand
Soils around drums or tanks: True
Soil contaminant: Volatile organic compounds
VOCs in soils pose the primary risk: False
Excavation acceptable: True
Site conditions could threaten: False
Exposed to erosion: False

RECOMMENDATIONS FOR HOT SPOTS (UNSATURATED MTL
AROUND LEAKY TANKS OR DRUMS)

GENERAL
o   503 Groundwater monitoring

IN SITU SOILS TREATMENTS
Either
o   305 Soil vapor extraction for VOCs
    Either
    o   306 Flare for VOCs
    Or
    o   308 Vapor phase carbon
Or
o   Deal with the principal threat (i.e. buried
    tanks or drums)

REMOVAL OPTIONS
o   201 Soil excavation

Figure 2. Example of CORA Expert System Output

Knowledge Bases. As shown in Figure 3, the second phase CORA expert system has four knowledge bases. The first knowledge base—CORA—is a small one of 15 rules that simply asks the user to specify the waste types for each contaminated area, then calls the second knowledge base—MAIN—to examine each waste type. MAIN has 492 rules that are grouped into the following categories: removal, treatment, containment, landfill, above-ground contamination, natural attenuation, and active restoration. The rules examine all contaminants specified by the user and try to recommend suitable RA technologies. The third knowledge base—LANDFILL—contains 71 rules and deals exclusively with landfill issues for by-products generated by treatment or containment. The fourth knowledge base—WATER—has 43 rules and deals with the treatment of liquids generated by the treatment or containment recommended by MAIN or LANDFILL.

To minimize the number of rules and thus reduce the RAM required to run the expert system, we took full advantage of the chaining capability of Level 5. Chaining capability is the ability of knowledge bases to call each other. For example, once the MAIN knowledge base determines that a landfill may be an option for a waste by-product, the LANDFILL knowledge base is called and the relevant facts are passed to it through a text file. The facts are then updated and passed back to the calling program. This eliminates redundant coding because each subsystem deals with only a small subset of the domain knowledge. The increased efficiency, however, is acquired at the cost of a loss of context knowledge by the expert system as a whole. Because each knowledge base deals with only a small set of facts, the user cannot restart the system at an arbitrary point (for example, during the execution of the LANDFILL knowledge base), change a previous answer, and restart the system from that point without losing the context of the consultation session.

Another problem we encountered with the Level 5 software was its erratic behavior when a small knowledge base such as CORA chains to a large knowledge base such as MAIN. Global facts that should be shared were not passed to MAIN. To work around this problem, we had to write these facts to a text file that is read by MAIN at start-up.

Rules and Confidence Factors. Facts that are known to the system are given a confidence level of either 0 (not true) or 100 (true). Facts that are not yet known to the system are assigned a confidence level of -1, and facts that are declared unknown are given a confidence level of -2. CORA was not structured to ask users for a degree of confidence in their answers. Each rule's conclusion is assigned either 0 if the rule fails or 100 if the rule succeeds. In that respect, CORA is not designed to deal with "fuzzy reasoning."

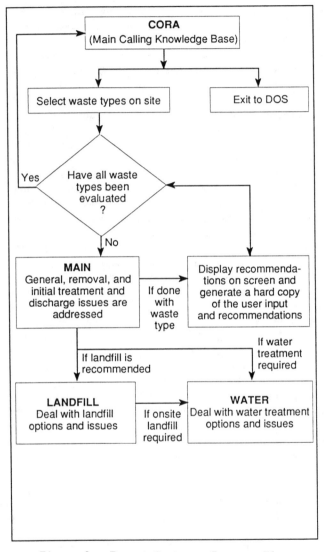

Figure 3. Expert System: Program Flow

There are, however, rules that check for unknown facts and try to pursue a failed line of reasoning if the line of reasoning is necessary for achieving the current goal.

To increase the robustness of the system, the knowledge engineers have implemented two ways of dealing with uncertainty reasoning. One method is to use multiple rules with the same conclusion. For example, one of the goals when dealing with groundwater is to determine the value of the hydraulic conductivity of the saturated zone. If the user indicates in response to a system question that this value is unknown, other rules with the same goal will ask the user to choose from a list of 12 attributes the one that describes the saturated zone in question (such as clay, gravel, sand). On the basis of this selection, the system assigns a default hydraulic conductivity value to the zone and proceeds to the next rule.

Another method of dealing with uncertainty reasoning is to check explicitly for facts that have a confidence level of -2. For example, if the contaminated area is groundwater, the user will be asked to select either natural attenuation or active restoration as a response action. Rules for checking the confidence level of the response action are in place in the knowledge bases. If the determination is that the fact has a value of -2, other rules are in place for determining this goal through a series of alternative questions.

Control Structure. The rules are invoked in a backward-chaining scheme that produces a depth-first search--that is, the inference engine tries to satisfy a goal by evaluating all rules that lead to that goal. Heuristic rules were used as much as possible to shorten the search. For example, the system may ask the user to select the desired response action, either containment or treatment of the wastes. If the user chooses containment, rules for treating the main wastes will not be activated. If by-products of the containment technologies require treatment, however, rules for treatment will still be explored.

Interface with External Programs. One drawback of Level 5 is that the tracing capabilities exist only in the production environment. Compiled versions of the knowledge bases do not allow users to see the line of reasoning or to change or look at facts processed during a consultation. So that users can keep a record of their consultation, additional user-interface and reporting capabilities were added via dBASE III+ programs that are activated from within the knowledge bases.

When the system receives a user response during a session, the system question and user answer are written to disk. When the system arrives at a recommendation, the recommendation is also written to disk. Such internal bookkeeping added greatly to the number of rules and the complexity of the knowledge bases.

At the end of the session, the dBASE III+ program is loaded into memory to gather and sort the data file. The program displays the system's recommendations on the screen and gives the user the option of having a printed record of the session. This printed record would include the system's questions, the user's answers, and the system's recommendations.

Testing and Validation. Both the first and second CORA prototypes were tested by in-house domain experts before being field tested. After each modification of the system, test cases were rerun to ensure that the system functioned correctly. In May 1987, the first prototype was field tested on 97 U. S. EPA Superfund sites likely to be FY 1989 RA candidates. For each site, CH2M HILL team members worked one-on-one with U. S. EPA RPMs and completed runs of the CORA expert system and the cost system. The second version of CORA was also tested by a selected group of EPA RPMs and CH2M HILL experts before it was released. As a result of the testing, additional explanatory information was added to the rules that might be confusing to users. The system was also modified to inform the users of intermediate recommendations.

Our two approaches to implementation presented an interesting contrast in rule modification and debugging during the testing cycle. We found that the approach in the first phase—writing rules that explicitly called each other—made debugging easy but rule modification difficult. Because each rule was linked to another, inserting or deleting rules meant that all possible rule links had to be examined and extended or truncated. The inverse was true in the second phase, where our approach of writing modular rules made rule insertion easy and debugging more complicated because the order in which rules were fired was not easily traceable. After the release of the second phase CORA model, EPA retained an outside consultant to conduct a validation study of the model. The study, conducted in January 1989, included a review of the decision rules and of the expert system's operational recommendations. The study concluded that the expert system is a "necessary and useful tool" (ICF, Inc. Performance Evaluation of Cost of Remedial Action (CORA) Model, prepared for the U.S. Environmental Protection Agency's Hazardous Site Control Division, January 13, 1989) and captures the decision process used in the Superfund remedial program.

Applications of the Expert System

The CORA expert and cost systems have been used in a variety of applications. The model was used to scope and help develop both EPA's and the Navy's remediation budgets for FY 1989, FY 1990, and FY 1991.

CORA was also used for regulatory support for the Resource Conservation and Recovery Act (RCRA) and for analyzing corrective-action strategies and costs for the RCRA Location Standards Rule. The model is being used to screen and evaluate technologies and remediation strategies for the Department of Defense.

The CORA model has also been used to screen technologies, develop alternatives, and estimate initial remediation costs. To date, more than 200 copies of the model have been distributed to federal and state agencies, foreign governments, environmental consultants, and industry.

Both the CORA expert system and the cost system were designed to allow revision and expansion. EPA funds have been appropriated for continued maintenance, enhancement, and incorporation of user feedback to reflect current regulatory policies, demonstrated technologies, and cost considerations.

Use of the CORA model is expected to continue. Future applications include:

o   Use by EPA and other federal agencies in evaluating remediation strategies and in developing fiscal outyear budgets

o   Use by federal agencies, states, industry, and environmental professionals in screening technologies and evaluating remedial alternatives

o   Use by states for total-program and site-specific remediation budgets and scoping

o   Use in anticipating choices of remedial technology and cost effects for regulatory-impact analyses of new environmental regulations

Summary

We believe CORA was a success because our emphasis for cost and time was mainly on conceptualization, formalization, and knowledge acquisition, as indicated by Figure 4. Despite the difficulties encountered in the Level 5 software, we believe it was an appropriate tool for developing the expert system because it required very little RAM to operate in comparison to most other shells on the market. The Level 5 software allowed us to meet our goal of developing a system that would run on a 640K microcomputer. In addition, the Level 5 programming language is relatively easy to learn, and the technical support for it was excellent.

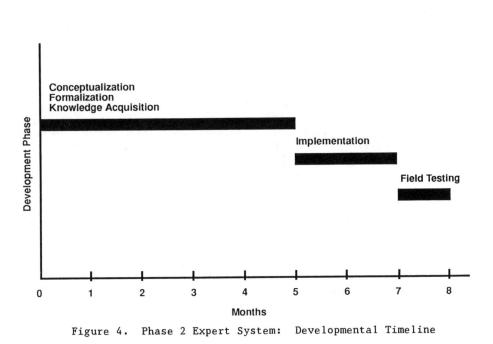

Figure 4. Phase 2 Expert System: Developmental Timeline

Acknowledgments

CORA was developed under EPA contract number 68-01-7090. The authors would especially like to thank Kirby Biggs and Russ Wyer of EPA's Hazardous Site Control Division for their help and support during the development of the model.

Literature Cited

1.   Brachman, Ronald J.; Levesque, Hector J. Readings in Knowledge Representation; Morgan Kaufmann Publishers, Inc.: Los Altos, California, 1985.

2.   Waterman, D. A. Guide to Expert Systems; Addison-Wesley: Reading, Massachusetts, 1986.

3.   Guidance on Remedial Actions for Contaminated Groundwater at Superfund Sites, U.S. Environmental Protection Agency, Hazardous Site Control Division, Office of Emergency and Remedial Response, 1988.

RECEIVED January 18, 1990

Chapter 13

Computerized System for Performing Risk Assessments for Chemical Constituents of Hazardous Waste

John L. Schaum[1], John J. Segna[1], John S. Young[2], Carol M. Benes[2], and Warren R. Muir[2]

[1]Office of Research and Development (RD-689), U.S. Environmental Protection Agency, 401 M Street, S.W., Washington, DC 20460
[2]The Hampshire Research Institute, 1800 Diagonal Road, Alexandria, VA 22309

The U.S. Environmental Protection Agency has sponsored the development of a software system to assist environmental personnel in conducting risk assessments at hazardous waste sites, and also in reviewing assessments generated by contractors for correspondence with EPA and state standards. This computerized system, called Risk*Assistant, combines a series of tools, including databases, expert exposure and risk values. These tools are provided in an IBM-PC format, with a user-friendly interface that allows a user to begin using the system quickly with little or no training.

As the process of regulating and controlling the release of potentially hazardous chemicals into the environment becomes a more complex task, the U.S. Environmental Protection Agency (EPA) has moved vigorously to a risk assessment / risk management / risk reduction framework for making regulatory decisions. Consequently, the EPA has undertaken a major effort to gain an understanding of the risk assessment process and to consider quality and consistency in risk assessments a major goal for the 1980s and 1990s. Part of the process in achieving this goal has been the development of EPA risk assessment guidelines. These guidelines include four areas concerning health effects, while the fifth deals with exposure assessment (1)

Components of Risk Assessment

The National Research Council (2) has divided the process of risk assessment into four components. These are:

(1) Hazard Identification - consists of a review of relevant biological and chemical information bearing on whether or not an agent may pose a specific hazard.

0097–6156/90/0431–0176$06.00/0

(2) Dose-Response Assessment - involves describing the quantitative relationship between the amount of exposure to a substance and the extent of toxic injury or disease.

(3) Human Exposure Evaluation - involves describing the nature and size of the population exposed to a substance and the magnitude and duration of their exposure. The evaluation could concern past, current, or future exposures.

(4) Risk Characterization - involves the integration of the data and analyses from the first three components to determine the likelihood that humans will experience any of the various forms of toxicity associated with a substance.

The U.S. EPA approach to risk assessments for toxic chemicals follows the format described by the NRC. Because Hazard Identification and Dose-Response Assessment for an agent do not depend upon specific local situations, EPA assumes that risk assessors evaluating specific sites will not conduct independent analyses in these areas but will instead rely on the results of peer-reviewed evaluations by qualified authorities in toxicology. EPA is assembling an agency-wide database of such authoritative assessments, the Integrated Risk Information System (IRIS).
Although the Hazard Identification and Dose-Response Assessment of an agent are generic, the risks that the agent poses are specific to a particular location and set of circumstances in which exposure to the agent occurs. In order to assess site-specific risks of an agent, the assessor must perform a specific Exposure Evaluation for the site, and combine it with information on the Hazards posed by the agent to yield a Risk Characterization for the site.
EPA risk assessment practice, (and the Risk*Assistant software) reflects this distinction between generic and site-specific information in the prediction of the risks associated with an agent at a specific site. The results of authoritative Hazard Identifications and Dose-Response Assessments are not expected to be re-evaluated by the risk assessors evaluating specific sites; these assessors are expected to concentrate on site-specific Exposure Evaluation and Risk Characterization.

Human Exposure Assessment. The assessment of human exposure involves the measurement or estimation of the number of people exposed and the magnitude, duration, and timing of their exposure. In some cases, it is straightforward to measure human exposure directly by measuring levels of the hazardous agent in the ambient environment. In most cases, however, detailed knowledge is required of the factors that control human exposure, including those factors which determine the behavior of the agent after its release into the environment. The following types of information are required for an exposure assessment:

(1) Quantities of an agent that are released and the location and timing of release.

(2) Factors controlling the fate of the agent in the environment after release, including factors controlling its movement, persistence, and degradation.

(3) Factors controlling human contact with the agent, including the size and distribution of the population, and activities that facilitate or prevent contact.

(4) Patterns of human intakes.

The amount of each type of information that is available varies greatly from case to case and may be difficult to predict accurately. Therefore, except in fortunate circumstances in which the behavior of an agent in the environment is unusually simple, uncertainties arising in exposure assessments can be significant.

The first two factors of the four listed above are addressed by a wide range of environmental fate and transport models that vary in theoretical and mathematical complexity and in the extent to which they have been validated. EPA has sponsored research and development of numerous such models; among the most successful implementations is the Personal Computer - Graphical Exposure Modeling System or PC-GEMS (3) developed under the sponsorship of the Office of Toxic Substances. Such models help the risk assessor to predict the number of people who may be exposed to chemicals leaving a site, and the environmental concentrations to which they might be exposed. They do not, however, address the crucial issues of the pattern of human activities that result in contact with the agent, and the amounts of the agent with which people may come in contact.

Such exposure information, rather than predicted or measured environmental concentrations, is needed to predict the risks associated with an agent. (Risk*Assistant was explicitly designed to provide risk assessors with assistance in this previously neglected area.)

Just as the models used to predict the transport and fate of chemicals in the environment are sensitive to numerous site-specific parameters such as average rainfall and soil types, the equations used to determine exposures for particular activity patterns are sensitive to demographic parameters. These include general population characteristics (e.g. age distributions), culturally-influenced factors (e.g. rates of fish and vegetable consumption), and location-specific factors (e.g. workplace exposure patterns are generally different from those in the home). EPA has recently published the results of its efforts to determine values for numerous parameters that are characteristic of the average population of the U.S. (4), but the risk assessor must adjust these parameters to fit the specific population she or he is evaluating.

Risk Characterization. The final step in risk assessment involves bringing together the information provided in the exposure assessment with information on the toxic hazards of an agent to determine risk. For agents which may be carcinogens, the carcinogenic potency (also called the slope factor) and Lifetime Average Daily Exposure (LADE) are used to derive an estimate of the probability that the specified exposure will increase cancer incidence over background rates (the actual equation is [1 - exp (-LADE x Slope)], which precludes risks greater than 1.0). For non-cancer toxic effects, standard practice is to compute a Hazard Index, which is the ratio of the Average Daily Exposure (ADE) to the Reference Dose (RfD). Hazard Indices less than 1.0 are assumed not to be associated with a significant risk of toxic effects.

While the risk characterization calculations themselves are straightforward the way in which the information is presented is important. For example, this step can be far more complex than indicated here, especially if problems associated with duration and/or timing of exposure are considered. Further, at many sites it is necessary for the assessor to consider possible interactions among several agents. The non-carcinogenic and the carcinogenic risk values, together with their associated estimates of uncertainty, are the final measures of the possibility of human injury or disease from a given exposure or range of exposures.

Standardizing Risk Assessment

The EPA relies on a risk assessment approach to make better regulatory decisions. However, risk assessment is a multi-disciplinary process potentially involving the efforts of analytical and environmental chemists, biologists, environmental engineers, statisticians, toxicologists, and others as appropriate (5). It is rarely possible to obtain extensive involvement from professionals in all of the relevant disciplines for any site-specific risk assessment. Staffing limitations mean that the risk assessor in the field must often make do with limited help from professionals in some relevant disciplines.

EPA is naturally interested in producing risk assessments of the highest possible quality. This requires that each assessment is sensitive to site-specific conditions, while using procedures that are consistent with those of other agency-sponsored assessments. Both inconsistency in approach and lack of technical quality (such as insensitivity to local conditions) represent problems in the risk assessment process. EPA is actively engaged in several efforts to maintain quality and consistency in risk assessment, in the face of distinctly limited professional resources.

The training of professional personnel in the risk assessment process is an ongoing activity at EPA. However, classroom training is only a partial answer to EPA's risk assessment dilemma. It lessens, but has not eliminated, the pressure of EPA to review technical reports with less experienced professional personnel. The development of a computer software system, Risk*Assistant, to assist environmental personnel in conducting risk assessments and reviewing assessments generated by contractors, represents a complementary approach to increasing the technical quality and consistency of risk assessments.

Risk*Assistant Capabilities

Software for risk assessment must address a variety of information manipulations, including data retrieval, categorical or rule-based decisions, and mathematical calculations. Such a range of manipulations are not fully addressed by any single programming approach. For example, many expert system shells are not optimal environments for performing extensive calculations, although they are far more convenient than programming languages as a means of encoding categorical decisions.

In view of the limitations of each of the common approaches to programming for addressing the range of problem types encountered in risk assessment, Risk*Assistant was designed as a modular software system. In other words, it is a series of separate programs, each of which addresses a particular component of risk assessment. All the programs are linked by a common "system shell," that ensures each program is provided with the information it needs from other programs and allows users to use as many of the programs as are needed with a minimum of effort. Because of its modular design, Risk*Assistant can readily be "customized" to address the specific analytical needs of any designated group of users. The current version has focused on the needs of EPA and state personnel who generate or review risk assessments of hazardous waste treatment, storage, and disposal facilities.

Components of Risk Assistant. There are three main components of the Risk*Assistant software system, each of which reflects a different approach to using the information contained in the system. Each segment makes use of the same family of program modules, but selects different subsets of these modules to address particular questions.

Main Analyses. The main analyses section of the software system includes those program modules that directly assist the user in either generating or reviewing exposure and risk assessments for hazardous waste sites. They include modules for Case Description, Exposure Assessment, and Risk Characterization, as well as a series of Superfund Checklist Modules.

Case Description. Obviously, any site-specific risk assessment requires information on the relevant characteristics of the site. The Case Description module allows the user of Risk*Assistant to enter or modify both the descriptive information regarding a site of potential chemical release and the sampling and analytical data (including quantitation limits) associated with that site. It also allows the user to select subsets of data associated with a site for use in any of the other Risk*Assistant analyses.

Exposure Assessment. As noted above, the Risk*Assistant software is intended to build on EPA's existing information base on environmental fate and transport modelling, extending it to risk-relevant exposure calculations. Accordingly, it does not incorporate mathematical models of the environmental transport and fate of chemicals, but takes as its starting point user-specified data on environmental concentrations of chemicals to which people might be exposed. The Additional Analyses discussed below, however, do include tools to assist the risk assessor in selecting appropriate transport models.

As noted above, the exposures actually experienced by people in an environment contaminated by chemicals depend not only the concentrations of the chemicals in environmental media but also on the specific details of the activities in which they engage. Factors such as inhalation rates, time spent in the home and outdoors, and the amounts of various types of foods consumed will have significant impact on exposures at any given environmental concentration of a chemical.

EPA has recently released the first volume containing the results of an effort to obtain the most reliable possible values for a variety of exposure parameters. This document, The Exposure Factors Handbook (4), covers 12 commonly considered exposure scenarios. The exposure assessment module of Risk*Assistant incorporates the algorithms for calculating exposures under each of these scenarios, for all environmental media for which the scenario is applicable. The user can select any or all of the exposure scenarios that are relevant to the environmental media that are contaminated at a site. Where more than one contaminated medium could influence a scenario, the user has the option of selecting the most appropriate medium.

As a further aid to the risk assessor, Risk*Assistant incorporates a database of the "average" and "reasonable worst-case" values for the parameters applicable to each of the exposure scenarios. The "average" value is automatically provided as a default. However, such "average" values may not be appropriate for particular locations or populations, and the user has the ability to substitute the "reasonable worst case" parameter value, or any other appropriate value for the specific population under consideration, for each parameter in each scenario. Such an approach provides for risk assessments that are responsive to the context of specific sites, yet retain a fundamental consistency of approach.

Sets of scenarios and parameters that apply to particular populations of interest to a user may be stored for future use. Exposure values are reported for each route of exposure (oral and inhalation routes are covered in the current version and the dermal route will be developed in the future) for each different scenario and for each route of exposure per each contaminated medium.

A key additional feature provided by the exposure assessment module is the ability to automatically calculate several indices of the uncertainty associated with the assessment. With a few keystrokes, the user can calculate the exposures associated with both "average" and "reasonable worst case" parameter values, as well as other combinations of parameter values. In addition, future versions will allow the user to examine the relative contribution to exposure of each of the scenarios selected for evaluation.

Risk Characterization.   Once a quantitative exposure assessment has been made, Risk*Assistant allows the user to automatically calculate lifetime excess cancer risk and/or a hazard index for toxic noncarcinogenic effects of chronic exposure for any agent included in the toxicity databases which currently include about 300 compounds. The appropriate hazard values (slope-potency factors and reference doses) for the relevant routes of exposure are automatically retrieved from the databases. The uncertainty calculations in the exposure assessment can also be retrieved to assess the range of risks associated with a given exposure situation.

Risk*Assistant yields a separate risk estimate for each chemical under consideration, and for each route of exposure to the chemical (oral, inhalation, or dermal). The user can readily combine these estimates to obtain an overall estimate of the risks associated with a site, but the software does not automatically provide such combined estimates, because to do so requires considerable judgement. The toxicity of any given chemical can vary qualitatively, as well as quantitatively, with the route of exposure; if so, it would be inappropriate to combine risk estimates from different exposure routes. The judgement of which chemicals will produce additive, less-than-additive, or more-than-additive toxic effects similarly requires more detailed knowledge of their toxic modes of action than is contained within this software.

Superfund Checklists. Under the Comprehensive Environmental Response, Compensation, and Liability Act (CERCLA or Superfund), much of EPA's responsibility consists of reviewing risk assessments generated by other parties. Recently, the guidance document that indicates appropriate procedures for performing risk assessments under Superfund (6) has been completely replaced (7). A series of automated check-lists are being incorporated into Risk*Assistant that will assist personnel in evaluating risk assessments generated by other parties for consistency with this new guidance. The initial modules include procedures for reviewing the collection and analysis of samples of environmental media that may be contaminated, and for reviewing the assessment of human exposures. Like the Exposure Assessment Module described above, the Exposure Checklist will automatically calculate the effects of alternative exposure scenario parameters on the resulting exposures.

Additional Analyses. Although the Main Analyses provided in Risk*Assistant cover the essential core of site-specific risk assessment, it is anticipated that risk assessors will also need assistance in other areas. The software system currently addresses three concerns: the need for setting priorities for site review based on minimal data, the frequent need to use models to predict the transport of chemicals from the site to populated areas, and the need to distinguish between probable and less likely conditions of exposure.

Quick*Risk. QUICK*RISK is a program designed for rapid, "back of the envelope" risk calculations, primarily for the purposes of screening or directing further research. The user need only provide a list of chemicals and concentrations (or estimates of the chemical concentrations) for air, water, fish, or soil. The system then uses a set of pre-specified assumptions to report on the risks of the chemicals at the specified concentrations from drinking water, inhalation, fish consumption, or soil ingestion. Both lifetime excess cancer risks and chronic non-carcinogenic (toxic) risks are evaluated, using "reasonable worst case" assumptions for the selected exposure scenarios. The system also estimates concentrations in each selected environmental medium that correspond to a one-in-a-million carcinogenic risk.

Model Selection Assistants. As noted earlier, a wide variety of models have been developed to predict the transport of contaminants in various environmental media. These models differ in the amount and type of data they require, in the nature and complexity of the underlying processes they reflect, and in the specific contaminants and sets of environmental circumstances they were designed to

evaluate. EPA has recently published two guidance documents to help
the exposure and risk assessor select mathematical models for use in
exposure assessments (8-9). These two documents cover surface water
and groundwater models and assist in matching models to the analytical
needs, available data, resources, and model experience of the user.

Expert systems planned for Risk*Assistant will incorporate the
logical structure and information from each of these two documents,
and use a series of questions regarding the site and the goals of the
modeling exercise to guide the user in selecting an appropriate
transport model. Anticipated future developments of these systems
will provide more extensive information to the user on the reasoning
employed to match models to a user's analytical needs and resources.

Exposure Pathways. An exposure pathway for a toxic chemical consists
of a means by which the chemical is released to the environment, its
transport over a short or long distance to an area in which people may
come in contact with it, which may or may not involve its movement
from one environmental medium to another, and its coming into contact
with the skin, digestive tract, or respiratory system of a person.

An expert system planned for Risk*Assistant addresses the
probability that various exposure pathways will be of concern at a
hazardous waste site. This module guides users in considering factors
that may increase or decrease the likelihood that a chemical will be
released from a unit such as a landfill or surface impoundment to any
environmental media. If release cannot be ruled out, factors
affecting transport to an area of potential exposure are considered.
If the presence of contamination in an area of potential exposure
cannot be discounted, potential exposure scenarios are reviewed. The
end result is a listing of potential exposure pathways that the user
may have to consider for the site.

Databases. The Database component of Risk*Assistant allows the user
to look up information directly in any of the databases, without using
the analytical programs. These databases are automatically called up
as necessary during Risk*Assistant analyses, but a user may simply
want to report specific information about a chemical or chemicals.
The databases contained in the system are described below.

Toxic Hazards. Either information on the carcinogenic potency of a
chemical (Slope Factor, Weight-of-Evidence Class designation),
estimates of non-carcinogenic toxic potential (Reference Dose,
Uncertainty and Modifying Factors, and Statement of Confidence), or
both, can be reported for several hundred chemicals. Values for both
inhalation and oral exposure are available. If the desired toxicity
value for a chemical has not yet been included in IRIS, an alternative
database extracted from the Health Effects Assessment Summary Tables
(HEAST) is automatically searched.

Chemical Properties. The physical-chemical properties of an agent
influence not only its fate and transport in the general environment
but also its behavior in micro-environments in which exposure may
occur. Thus, the volatility and water solubility of a chemical will
influence its tendency to volatilize from domestic water into

household air. Among the more important physical-chemical properties included in the databases are:

(1) Molecular Weight
(2) Vapor Pressure
(3) Water Solubility
(4) Henry's Law Constant
(5) Octanol-Water Partition Coefficient
(6) Organic Carbon Partition Coefficient
(7) Bioconcentration Factor
(8) Diffusivity in Water
(9) Diffusivity in Air
(10) Melting Point
(11) Boiling Point

Regulatory and Advisory Standards. In some cases, rather than performing a detailed risk assessment for a specific situation, a user may wish to rely upon the risk assessments that underlie various published regulatory and advisory standards. In other cases (such as at Superfund sites), the user would want to know the relationship between observed concentrations and regulatory/advisory standards even if she or he intended to perform an independent risk assessment. Risk*Assistant contains databases of standards and advisory values developed by the Federal government, State governments, and other bodies, such as:

Federal Standards:

(1) Maximum Contaminant Levels and Maximum Contaminant Level Goals specified under the Safe Drinking Water Act

(2) Reportable Quantities for environmental discharge specified under the Clean Water Act and Comprehensive Environmental Response, Compensation, and Liability (Superfund) Act

(3) National Ambient Air Quality Standards specified under the Clean Air Act

(4) Permissible Exposure Limits for workroom air specified under the Occupational Safety and Health Act

State Standards:

(1) State Water Quality Standards

(2) New Jersey Maximum Contaminant Levels for Drinking Water

Advisory values:

(1) Drinking Water Health Advisories specified under the Safe Drinking Water Act

(2) Ambient Water Quality Criteria specified under the Clean Water Act

(3) Threshold Limit Values for workroom air promulgated by the American Conference of Governmental Industrial Hygienists

State water quality standards will frequently be "applicable" or "relevant and appropriate" requirements for a Superfund site. For most states and chemicals, the particular standards that apply to a water body depend upon the designation of the water body as a member of a particular class, with the set of classes varying among states. Risk*Assistant contains an automated procedure to help the user in selecting the most appropriate water body classifications for her or his particular site.

User Interface. The object of software systems for risk assessment is to make the process of generating or reviewing risk assessments easier for system users. Software that is not easy to learn or use, or that requires constant reference to manuals or other support documents, will do little to help risk assessors faced with a demanding workload. Accordingly, a key emphasis in the development of Risk*Assistant is that the software be usable by persons with little or no computer experience, without the need for training, and that the majority of a user's questions can be answered in the software, without the need to refer to manuals. The development of technical manuals is an important part of the Risk*Assistant effort, but the user should not need to refer to these manuals frequently.

To avoid the need for users to rely on manuals, several features have been added to the user interface for Risk*Assistant (i.e., the menus by which the user enters and retrieves information). These are described briefly below. In addition, an effort has been made reduce the amount of typing required of the user to a bare minimum and to make the keystrokes that activate the system as obvious as possible.

Help Screens. Each menu in Risk*Assistant has one or more associated HELP screens, accessible by pressing a function key. The HELP screen explains how to select an item from a menu, change a default value, or enter data.

Explanation Screens. In addition to HELP screens that instruct the user how to do something, EXPLANATION screens tell the user why she or he is being asked to make a choice or enter data. A brief explanation of the relevance of a particular choice to the overall risk assessment process is provided; if appropriate, alternative values for data entries are given.

Reference Screens. Whenever Risk*Assistant provides a default value, a REFERENCE screen can be called up to provide a citation for the literature source from which the value was taken.

Notepad. Each menu in Risk*Assistant has an associated Notepad screen. By pressing a function key, the user can call up the notepad for the menu, and use it to explain important features of his or her assessment, or the reasons underlying the selection of particular values.

Novice User Pathway. The user of Risk*Assistant has the option, at any point, of using the system in either an "experienced user", or a "novice user" mode. A function key toggles between the two modes. In the "novice user" mode, each key menu is preceded by one or more screens explaining its place in the risk assessment process and indicating how to make an appropriate response.

Design Considerations

To meet EPA's goal of improving the consistency and quality of risk assessments, the software must not only be easy to use, but must also generate results that can be relied upon in a regulatory context. Users must have confidence in the accuracy of the information contained in the system, whether database parameters or rules and algorithms. Further, the logic used by the software in reaching any conclusion must be explicit. A "black box" approach, even if infallible, would be of very limited utility in the regulatory environment.

Authoritative Databases. Efforts to ensure the accuracy of the information in Risk*Assistant databases involve both the selection of data sources and Quality Assurance procedures for data entry. As noted below, an effort was made to locate the most authoritative source for each database. Database entries are repeatedly checked against original sources. Finally, the user is supplied with a citation of the original literature source, and so is able to confirm database contents if necessary.

Toxic Hazards and Exposure Parameters. Supplying EPA users with authoritative information on toxic and carcinogenic hazards has been relatively straightforward.

(1) Integrated Risk Information System (IRIS) - Wherever possible, Risk*Assistant obtains information on chemical hazards from the EPA's Integrated Risk Information System (IRIS). All information on chemical hazards contained in IRIS is subject to rigorous peer review and represents and Agency-wide standard value (10).

(2) Health Effects Assessment Summary Tables (HEAST) - Because of the rigorous peer-review process, IRIS does not yet contain information on all chemicals of potential interest to users of Risk*Assistant, or on all aspects of the toxicity of the chemicals that it does include. Accordingly, for chemicals that are not covered by IRIS, toxic hazard data from the HEAST are provided. These data do not have the Agency-wide approval of IRIS data, but have been reviewed by EPA.

Regulatory Standards. Authoritative database sources for regulatory standards are, like toxic hazard sources, easy to come by. Both the federal and state governments provide published versions of such standards that are reliable sources for database entry. The key issue for these databases is checking for accurate entry, and periodic review for currency.

Chemical Properties. Because no single source contains information on all relevant properties, information from a variety of sources is contained in these databases. Accordingly, all entries are referenced to the original source of the information.

Specialized Databases. Some potential users of Risk*Assistant may have developed independent estimates of toxic risk associated with chemicals, particularly for chemicals not included in EPA toxic hazard databases. Such users (such as State governments), who typically will have access to considerable toxicological expertise may wish to use these hazard values in Risk*Assistant analyses. Such specialized hazard databases can readily be incorporated into the system.

Referenced Calculations. Exposure assessment involves numerous calculations, covering both cross-media transfers of chemicals and the derivation of exposures from concentrations and scenario-specific parameters. In general terms, such calculations can be viewed as the limiting case (in simplicity) of either theoretical or empirical models. All calculations in Risk*Assistant for deriving exposures from concentrations in an appropriate medium (e.g. inhalation exposures from air concentrations) are obtained from the Exposure Factors Handbook (4). Equations for evaluating cross-media transfer for particular exposure scenarios (e.g. volatilization from domestic water to household air) are obtained from literature sources. For such equations, the original reference is provided for the user.

Expert Systems as "Automated Documents." Sometimes a user needs to make a decision based upon the consideration of categorical or qualitative information, rather than calculations involving numerical parameters. Expert Systems comprise several types of computer programs that address this type of information. Risk*Assistant will contain a variety of small expert systems that assist users in different aspects of the risk assessment process.

Expert systems sometimes represent an attempt to incorporate the views and judgments of a particular expert in a given discipline. This raises issues concerning the procedures used to identify the expert on whom to base the system and the possibility that other experts of equal competence would reach alternative conclusions, that can limit the utility of an expert system for use in a regulatory context.

Expert systems can also be "delphic" in their operation, reaching conclusions from antecedents by processes that are non-obvious to the user. Sometimes these processes are not even obvious to the expert and knowledge engineer who developed the system, but are derived by expert system software from a series of specific situations presented to the expert for evaluation.

For the regulatory context in which EPA evaluates hazardous waste sites, neither reliance on the simulated judgement of a single individual nor a "delphic" decision process are acceptable. Thus for purposes of this system the decision was made to use expert systems technology to incorporate explicit procedures from regulatory or guidance documents. Such documents have already been reviewed and deemed appropriate before publication, and by their nature they include explicit decision criteria. The incorporation of regulations and guidance documents into expert systems is the approach taken by Risk*Assistant.

A key feature of ongoing expert systems research in this project is developing expert system reports that clearly document the reasoning used to proceed from antecedent conditions to a conclusion. Having such information is necessary in the regulatory context, where decisions based on "delphic" processes are likely to be challenged as "arbitrary and capricious."

Applications of Risk*Assistant

As noted above, the programs and databases contained in Risk*Assistant are designed to serve both experts and novices in the exposure and risk assessment process, with the recognition that individuals with expertise in one discipline may be charged with the responsibility of conducting or reviewing exposure and risk assessments that incorporate data from many different disciplines. Potential users include regional EPA staff and their counterparts in state and local environmental agencies and other individuals (private, corporate) concerned with evaluating the health risks posed by hazardous waste sites.

Generation of Risk Assessments. Much of the risk assessment workload faced by EPA regional personnel and state personnel does not involve the development of detailed risk assessments that attempt to be as accurate as possible in characterizing particular situations. These professionals need to use relatively simple risk assessments to set risk management priorities. They need to be able to decide which situations require immediate attention, which can be deferred for later action because they pose little risk even under worst-case conditions, and which require more detailed evaluation. Risk*Assistant can help to make such decisions in a consistent manner, while maintaining sensitivity to important features of particular situations.

Screening with QUICK*RISK. For users who need to set assessment priorities for a large number of sites, QUICK*RISK provides a means to accomplish the task rapidly. The user need only specify probable contaminant concentrations in environmental media, and QUICK*RISK applies appropriately conservative exposure assumptions. Although only one scenario is considered for any medium, further detail is probably not needed for the initial selection of site priorities, and QUICK*RISK provides for consistent evaluation. Assuming that concentration estimates were available, several hundred sites could be evaluated in a single day.

QUICK*RISK also provides users with the ability to respond quickly to requests for risk information regarding poorly characterized situations. For example, a telephone inquiry about potential risks associated with water contamination could be answered in a matter of seconds.

More Detailed Analyses. When dealing with a more limited number of sites (dozens, rather than hundreds), Risk*Assistant allows the user to take specific site conditions into account while maintaining a consistent approach to risk assessment. The assessments can be produced quickly (an experienced user could perform several in a day), yet still contain enough to provide confidence in the decision made and a sound beginning for a full-fledged risk assessment if one is required.

Summary of Site Information. The case description section of Risk*Assistant provides a sufficient level of descriptive detail to enable a user to form a clear conceptual model of the site and to present relevant information concisely. It also enables a summary evaluation of the adequacy of the data base on a site.

Identification of Pathways. The expert system will guide a user to consider the variety of exposure pathways that should be considered for a site, and thus to perform a qualitative risk assessment. It helps to ensure that significant current or potential exposure pathways are not ignored.

Assistance in Modeling. Risk*Assistant does not perform transport modeling, because other EPA-developed systems already provide these capabilities. It does, however, assist the user in selecting the appropriate transport models for a site, if modeling proves to be appropriate. An effort is underway to facilitate the automatic transfer of information between Risk*Assistant and EPA modeling software, such as PC-GEMS (3).

Calculation of Exposure. A great advantage of Risk*Assistant is that it allows users to consider a wide range of factors that will influence quantitative exposure estimates (e.g., specific exposure factors for different populations, pathways, and scenarios) with a minimum of effort. Thus, a user can rapidly produce alternative exposure evaluations, including best estimates, reasonable worst-case exposure estimates, and worst case exposure estimates. Moreover, the system provides the user with information on the degree of uncertainty contributed by various assumptions used in the analysis, which can guide the user's future data collection efforts so that they result in maximum reduction of uncertainty.

Calculation of Lifetime Excess Cancer Risk and Hazard Indices for Chronic Exposure. For each exposure estimate, it is extremely easy to generate corresponding risk estimates. Thus, the user can quickly specify the full range of risks that may reasonably be attributed to a site. If reasonable worst-case or worst-case assumptions indicate minimal risk, it may be possible to defer the action until more pressing problems are addressed. Alternately, the Risk*Assistant analyses may indicate current or potential risks that should be immediately ameliorated. When a wide range of risks may apply to a site, it may be important to conduct additional studies to reduce the uncertainties associated with a site.

Reviews of Risk Assessment Generated by Others. In many cases, it will be important for EPA and state staff to review risk assessments generated by third parties (e.g., contractors or potentially responsible parties). Risk*Assistant facilitates such reviews, by providing data on standard procedures and assumptions as well as automated menus for comparing values actually used to standard values.

When a risk assessment under review deviates from the use of standard factors, there may in many cases be a valid reason for so doing; the assessment may reflect particular conditions at or near a specific site. The ability to annotate findings in Risk*Assistant provides a ready means for reviewers to indicate the importance and/or validity of deviations from standard procedures.

Sampling/Analysis Checklist. This module of Risk*Assistant, developed specifically for the review of Superfund risk assessments, prompts the user to consider key issues in quality assurance and quality control for sampling conducted at a waste site. It provides a concise summary of whether the standards developed under EPA's Superfund program have been met.

Quantitation and Detection Limits Reviews. Because the quality of laboratory analyses of chemical contamination is critical to the validity of any conclusions about risk, EPA initiated the Contract Laboratory Program (CLP) to ensure consistent, high quality analyses. By comparing reported quantitation limits for samples at a site to quantitation limits specified by CLP, Risk*Assistant will provide a report on a key feature of analytical quality sensitivity.

Comparison of Toxic Hazard Values with Standard Values. For the derivation of valid risk estimates, it is crucial that current hazard estimates from authoritative sources are employed in risk calculations. Risk*Assistant enables a reviewer to easily compare the values used in a particular assessment with standard values from IRIS or HEAST.

Exposure Assessment Review. Because so much of the difference among alternative risk assessments reflects the use of different assumptions about exposure, Risk*Assistant provides an explicit program for comparing exposure assessments to standard procedures. This program is an adaptation of the one used to generate exposure assessments, which was designed explicitly for reviewing such assessments, it considers several key components of the exposure assessment process, including:

(1)   Data Selection - Were average or maximum concentrations used (or were both used?). Did the averaging process use arithmetic or geometric means? Were any chemicals excluded from the analysis, and what was the rationale for excluding them?

(2)   Pathways Considered - Were any significant current or potential exposure pathways excluded from analysis? What is the consequence of including these pathways on total exposures? Were any inappropriate pathways considered? How much do these pathways contribute to the total exposure?

(3)   Scenarios and Routes of Exposure - Were all appropriate scenarios considered? For example, if domestic water

represents an exposure pathway, were bathing and contamination of household air by volatile chemicals considered as well as direct consumption? Were all relevant routes of exposure (oral, dermal, inhalation) considered?

(4)   Selection of Parameter Values - For the scenarios evaluated, were standard factors employed? Were these average values or reasonable worst-case values? If alternative values were used, what was the justification? Was the reasoning correct?

(5)   Results of Using Alternative Values - How do the exposure values obtained from the exposure assessment actually performed compare to values obtained using standard assumptions? How much uncertainty is associated with the exposure assessment?

Communication. In addition to its use for analyses, Risk*Assistant produces several standard types of reports that allow users to synthesize and summarize information on risk in a consistent format. Any assumptions that are used in an analysis are noted in these reports. These reports enable a user to effectively communicate key site information with other personnel, including risk management decision-makers and outside risk assessment consultants.

Record Keeping. Risk*Assistant will store and retrieve data entered for every exposure and risk assessment that has been performed and will automatically transfer data from one program to another, eliminating the need for repetitive data entry. It is easy to retrieve and review information from past assessments and to conduct multiple assessments on the same set of data with differing assumptions. Risk*Assistant's electronic notepad stores any annotations concerning an assessment.

Reference. Risk*Assistant is a ready reference source that reduces the necessity of flipping through numerous reference books or searching several external databases.

Literature Cited

1. The Risk Assessment Guidelines of 1986, U.S. Environmental Protection Agency, Office of Health and Environmental Assessment: Washington, DC, EPA/600/8-87/045, August 1987.

2. Risk Assessment in the Federal Government: Managing the Process. National Research Council (NRC), National Academy Press: Washington, DC, 1983.

3. PC-GEMS Users Guide, U.S. Environmental Protection Agency, Office of Toxic Substances, Exposure Evaluation Division: Washington, DC, February 29, 1988.

4. Exposure Factors Handbook, U.S. Environmental Protection Agency, Office of Health and Environmental Assessment: Washington, DC, EPA/600/8-89/043, March 1989.

5. Proposed Guidelines for Exposure-Related Measurements -- Request
    for Comments; Notice, Federal Register 53(48830-48853), December
    2, 1988.

6. Superfund Public Health Evaluation Manual, U.S. Environmental
    Protection Agency: Washington, DC, 1986.

7. Risk Assessment Guidance for Superfund: Human Health Evaluation
    Manual Part A. U.S. Environmental Protection Agency, Office of
    Solid Waste and Emergency Response: Washington, DC 20460. OSWER
    Directive 9285.701a, July, 1989.

8. Selection Criteria for Mathematical Models Used in Exposure
    Assessments: Surface Water Models, U.S. Environmental Protection
    Agency, Office of Health and Environmental Assessment: Washington,
    DC, EPA/600/888/042, July 1987.

9. Selection Criteria for Mathematical Models Used in Exposure
    Assessments: Ground Water Models, U.S. Environmental Protection
    Agency, Office of Health and Environmental Assessment: Washington,
    DC, EPA/600/888/075, August 1988.

10. Integrated Risk Information System Support Document, Vol I, II,
    U.S. Environmental Protection Agency, Office of Health &
    Environmental Assessment: Washington, DC, EPA/600/8-86/032a, March
    1987.

RECEIVED February 7, 1990

Chapter 14

Remedial Action Priority and Multimedia Environmental Pollutant Assessment Systems

J. G. Droppo, Jr. and B. L. Hoopes

Pacific Northwest Laboratory, P.O. Box 999, Richland, WA 99352

The Remedial Action Priority System (RAPS) and Multimedia Environmental Pollutant Assessment System (MEPAS) are different names for an objective exposure pathway evaluation system developed by Pacific Northwest Laboratory to rank chemical and radioactive releases according to their potential human health impacts. Constituent migration and impact are simulated using air, groundwater, overland, surface water, and exposure components based on standard assessment principles and techniques. A shell allows interactive description of the environmental problem to be evaluated, defines required data in the form of problem-specific worksheets, and allows data input. The assessment methodology uses an extensive constituent database as a consistent source of chemical, physical, and health-related parameters.

Pacific Northwest Laboratory has developed health impact assessment systems, the Remedial Action Priority System (RAPS) and the Multimedia Environmental Pollutant Assessment System (MEPAS), for the U.S. Department of Energy (DOE) to evaluate the relative importance of environmental problems. RAPS, which was developed first, applies to releases from inactive waste sites. MEPAS, the most recent version of the system, allows consideration of releases from both active and inactive sites. MEPAS differs from RAPS mainly in terms of the types of emission options. Although MEPAS retains the documented framework of RAPS (1), several enhancements have been added to the transport and exposure components (2).

The purpose of this paper is to describe MEPAS, a computer-based methodology for health impact estimation developed to support DOE site prioritization. MEPAS takes a physics-based approach based on characterization of exposures resulting from source-to-receptor transport at DOE sites. It is currently being used in DOE's Environmental Survey effort aimed at identification and prioritization of DOE sites. A Preliminary Summary Report has been issued for major DOE production sites (3), and a Final Summary Report for all DOE sites will be available in 1990.

Background

Making a comprehensive assessment of releases of hazardous materials
to the environment can be a difficult task. The actual and/or potential environmental issues must be defined in terms of releases, transport pathways, and exposure scenarios. An assessment approach must
then be selected. Options include making measurements, simulating
impact with environmental models, or employing a combination of measurements and modeling. The combination approach has the advantage of
allowing the use of measured values to calibrate the spatial and tem-
poral exposure patterns generated with models.
    Considering the large number of sites and diverse environmental
issues identified by DOE's Environmental Survey (3), DOE chose the
combination approach and selected a health impact modeling strategy as
an objective basis for ranking environmental issues. The goal was to
develop a system consistent with currently accepted assessment methods
that would have general applicability to the wide range of potential
problems. That system is MEPAS.
    The individual components of MEPAS are not based on standard, gen-
erally accepted techniques and methods such as those given by the
U.S. Environmental Protection Agency (EPA) (4). Rather, MEPAS is a
fully integrated physics-based exposure assessment system with a user-
friendly shell designed to allow application to a large number of environmental issues. This system has applications beyond site ranking in
site-specific remedial investigation and endangerment assessment
processes.
    The MEPAS methodology allows comprehensive site-specific eval-
uations of a large number of problems occurring in almost any environmental setting. Although similar systems are available, these
systems are limited to the local situation for which the system was
developed. MEPAS represents a composite of these systems. Although
these systems may provide better characterization of certain local
processes, MEPAS provides sufficient site-specific characterization of
local conditions for screening environmental issues.
    MEPAS allows the user to prioritize hazardous, radioactive, and
mixed-waste sites, based on their potential hazard to public health.
MEPAS is applicable to a wide range of environmental management and
regulatory conditions including inactive sites under the Comprehensive
Environmental Response, Compensation, and Liability Act (CERCLA), and
active sites releasing air and water contaminants regulated under other
statutory acts.
    MEPAS uses site and regional data to estimate long-term potential
health impacts. Average concentrations resulting in exposures can be
directly input if they are available from site monitoring or modeling
efforts. These average concentrations may alternatively be computed
using MEPAS emission and environmental transport modules. The principal output is a population impact index. Other outputs include
maximum individual impact, computed concentration fields, and constituent pathway scoring information.
    MEPAS uses mathematical algorithms and a coupled pathways analy-
sis to predict the potential for contaminant migration from a waste
site to important environmental receptors. Groundwater, overland, surface water (e.g., rivers and wetlands), and atmospheric pathways are
considered. Using the contaminant transport predictions, computed

exposure assessments based on direct contaminant ingestion and inhalation and on indirect intake through food production are performed for the identified receptors. The relative health impacts associated with the sites are then calculated.

MEPAS is not an expert system, because all logic paths and input are fixed. However, MEPAS does contain elements of an expert knowledge-based system and is of interest in terms of a framework for evaluating environmental concerns. MEPAS is one of the few currently available fully integrated systems that accomplishes the interactive definition, organization, and assessment of complex environmental problems.

Although MEPAS applications are often data intensive, the MEPAS documentation provides both application guidance (5) and definitions and sources of information for all input parameters (6). When available, site-specific data are used. Alternative procedures are provided for estimating emission, transport, and exposure parameter values for sites using local, regional, state, and federal information sources.

MEPAS was tested by EPA as part of a model comparison effort at Superfund sites (4, 7). An independent assessment was conducted by Morris and Meinhold (8). Testing efforts by the model developers are documented in Whelan et al. (9, 10). Additional information on MEPAS applications is given in Droppo and Buck (11), Hartz and Whelan (12), Poston and Strenge (13), and Buck and Aiken (14).

MEPAS Methodology

The MEPAS methodology is composed of the computer models and the supporting documentation. The computer components are completely integrated in a single user-friendly system referred to as the MEPAS shell (15). The application documentation has been structured to directly correspond to the operation of the MEPAS shell (5, 6). The transport and exposure components have been individually tested using monitored values at DOE sites (9, 10).

The migration and fate of contaminants in each transport pathway can be simulated using MEPAS components. The transport pathways are systematically integrated with an exposure assessment component that considers the type, time, and duration of exposure and the location and size of the population exposed. These various pathways and their interactions as considered by MEPAS are discussed in Droppo et al. (5).

A MEPAS application uses two sources of data: user input and constituent database. The user inputs site and regional data to define the nature of the issue, source term, transport pathway, and exposure scenarios. To help ensure consistency for a large number of applications, a constituent database was developed that contains chemical, physical, environmental, exposure, and toxicity data for each constituent. The constituent database used for the Environmental Survey is documented by Strenge and Peterson (16). This database currently has entries for 397 constituents; new constituents are added as needed.

MEPAS Structure

One difficulty in conducting a large number of applications is organizing the applications and their data. A standardized approach is
needed to allow a consistent definition of environmental problems.
For example, a formal classification of environmental problems can
provide a framework for data organization. Although the approach was
largely dictated by the Environmental Survey's classification, the
MEPAS structure should be applicable to environmental problems in
general.
    MEPAS contains a standard nomenclature for ease of organizing
environmental applications. This nomenclature includes the terms
"facility," "ranking unit," and "release unit," which have the following specific meanings within the MEPAS framework.

Facility. The term "facility" refers to a logical grouping of environmental problems. This grouping was mainly organizational (i.e.,
all operations at regional facilities) in the Environmental Survey,
although groupings by other geographic or political divisions could
also be used. The MEPAS facility is normally equivalent to the term
"site" under CERCLA and to the term "facility" under the Resource
Conservation and Recovery Act (RCRA), which includes containment
designs as part of the system under consideration.

Ranking Unit. "Ranking unit" refers to an environmental issue at a
facility. The definition of a ranking unit is derived directly from
its use in the Environmental Survey. A ranking unit is a composite of
similar and related environmental problems located in approximately the
same geographic location. There can be, and often are, multiple ranking units at each DOE site. Each ranking unit may have multiple exposure modes (e.g., multiple release locations, different release
methods, different pathways).

Release Unit. "Release unit" refers to the location in space and time
where a contaminant may be released into the environment.
    Figure 1 illustrates the relationship between a facility, ranking
unit, and release unit. The facility has a number of potential environmental problems: tanks, a building vent, and a landfill. The tanks
are shown as a ranking unit (environmental issue) associated with the
facility. The leak from a single tank illustrates a release unit subdivision of the ranking unit.
    These terms are sufficient for applications up to the point of
release of a hazardous material into the environment. For each
release, a "scenario" is defined to describe both the environmental
transport and the exposure routes.
    The terms "site" and "waste site" were not given specific mean-
ings because these terms are used to refer to several environmental
problem attributes. Also, the designation of a waste site refers only
to inactive sites, whereby MEPAS is designed to apply to environmental
releases from both active operations and inactive waste sites.

MEPAS Shell

The MEPAS shell was developed to allow expedient application of MEPAS to a large number of environmental issues. The MEPAS shell

- helps the user make specific applications and provides an overall database structure for making a large number of applications

- allows the user to interactively define a problem and approach for analysis of the problem

- creates problem-specific worksheets

- automates data entry, error checking, and documentation

- allows the user to make transport and exposure computations

- provides easy access and manipulation of site data and files for assessing potential health impacts from environmental releases of hazardous materials.

The transport and exposure components ($\underline{1}$, $\underline{2}$) can be run independently from the MEPAS shell. However, the MEPAS shell greatly improves the ease and accuracy of analyzing a large number of environmental problems by automating file creations, making range checks on inputs, and organizing required data.

The MEPAS shell provides user-friendly operation of problem definition, data entry, file creation, and execution of environmental simulation codes ($\underline{17}$). An IBM PC (International Business Machines Corporation, Boca Raton, FL) (or 100% compatible) with 640K RAM, a 20-Mb hard disk, and a 132-column printer are required. A math co-processor is not required, but it will greatly improve model performance. The data capture and storage programs are written in compiled dBASE III Plus (Ashton-Tate, Torrance, CA). The MEPAS transport and exposure models are written in FORTRAN, and intermediate linkage programs are written in 'C.' Installation requires approximately 2 Mb of disk space. This is a file-based application; all data are stored and exchanged between major components by file input/output.

Figure 2 is the main menu of the MEPAS shell. The menu options provide access to the following components:

- Constituent Library. The user can view the physical properties of radionuclides and chemicals used in the MEPAS model.

- Utility Routines. Various utilities allow the user to re-index database files, define the printer, and set up a color monitor.

- Create/Update/Examine. All data entry, edit, and review functions take place under this broad heading.

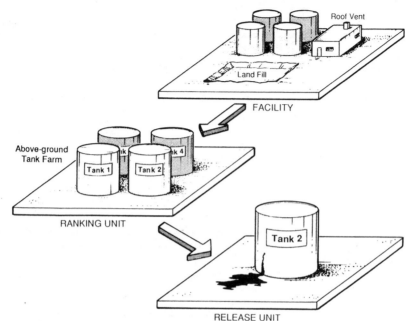

FIGURE 1.   Relationship Between a Facility, Ranking Unit, and Release Unit

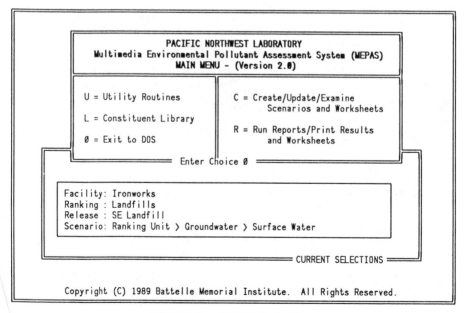

FIGURE 2.   MEPAS Main Menu

• Run/Print Reports/Templates. This selection enables the user to
specify run-time parameters, create run files, and invoke the
MEPAS model. Similarly, all printing is done from this menu
option, including generation of input data lists and intermediate
and final reports.

For the purposes of this overview, we will describe only the major
components and features of the MEPAS shell.

Constituent Library. From the main menu, the user has a view-only window to the constituent library. The constituent can be selected by
entering one or more characters or by moving through the database using
motion keys such as PgUp or right arrow. The view includes a tabular
list of the properties of the constituent required by the exposure
model. Properties of the constituents required by the model include
physical properties, such as vapor pressure, Henry's law constant, and
molecular weight; exposure data, such as reference dose, cancer potency, and bioaccumulation factor; and typical values for site data,
such as KOW, KOC, and K_ds. For the DOE application, the users could
not alter the contents of this library so that the same constituent
library would be used in all applications. These constituent library
parameter values and their sources are documented in Strenge and
Petersen (16). Parameters defined by the EPA were used as a primary
source of information.

Defining a Case for Analysis. The screen that appears after selection
of the Create/Update/Examine option from the main menu reflects the
structure for definition of cases for analysis by MEPAS. Descriptive
names are given to each subdivision of environmental problems. The
names of current selections appear next to the corresponding component,
which appears in a pattern on the screen illustrated in Figure 3. Each
box can be interactively selected (in sequence from top to bottom) to
access data entry screens that occur as sublevels of the boxes. These
data entry screens associated with each of the selection boxes shown
in Figure 3 allow the progressive definition of the case to be
analyzed.

   As noted above, the structure for definition of cases for analysis by MEPAS starts with the facility. A facility may have one or more
environmental issues (ranking units). The environmental issues may be
composed of one or more releases into the environment (release units).
These releases may then travel and impact people in one or more ways
(scenarios). To start the sequence of defining environmental problems,
the user positions the selector bar on the FACILITY designation, and
presses RETURN. In the next screens that appear, the user either
defines a new facility or selects a previously defined facility. For
a new facility, a duplicate of the entire database for a previously
defined facility may be used. The definition of a facility involves
facility location descriptors and relevant users and reviewers names.

   The next step is to give ranking unit names to the environmental
problems to be analyzed. After returning from the FACILITY selection,
the shell will, by default, move the selector bar to the RANKING UNIT
position to reinforce the concept of logical progression that is implemented throughout the shell. The user can either select from a list

of ranking units already prepared for the parent facility or add one. A new ranking unit can be based on duplication of data for a previously defined ranking unit. A data entry screen with location information needs to be completed for each new ranking unit. To define a release unit, the user positions the selector bar on RELEASE UNIT and presses RETURN. After returning from the RANKING UNIT selection, the shell will, by default, move the selector bar to this position. As with the ranking unit, the user can either select from a list of ranking units already prepared for the parent facility or add one.

The data entry screens for a release unit include the identification of constituents of concern. The user selects a maximum of 20 constituents from the constituent chemical properties database, including both parent and decay products. Fast interactive constituent search options are provided with simple on-screen constituent selection. To help in this selection process, alternative constituent names are listed. In addition, the user can access screens listing the properties for each constituent.

The SCENARIO selection activates screens for definition of the transport and exposure scenario to be used in evaluation of the environmental problems. The user can select from a list of scenarios already set up for the current ranking unit or choose to add one. To add a scenario, the user must choose the pathway that most closely represents the actual problem and then select those waste unit constituents that are transported by this particular pathway; add or select receptors; and match the receptors with exposure routes, such as ingestion, bathing, and direct contact.

After facility, ranking unit, release unit, and scenario data input are complete, the user has defined the major elements of the environmental problem to be analyzed. The user can then proceed to screens for entry of data needed to evaluate this problem.

Problem-Specific Data Templates. The required problem-specific data have been grouped logically and formatted into data entry templates. The first template is an interactive "control option" template used to define the structure and contents of the remaining templates.

Once the control option template is completed, a list is displayed of all templates required for the specific problem selected. The shell at this point has defined the data required to analyze a problem.

The user can either print copies of blank worksheets or start interactively entering the required data. Completed worksheets also can be printed. In the Environmental Survey, these data sheets are being used for data accuracy site reviews and, upon completion, as file copies.

Parameter data entries on these templates are coupled with reference number entries. In situations where data source referencing is required, the reference system allows definition of the source for each parameter value. This feature is activated by a function key and allows the user to enter short references or footnotes. These references can be created, modified, printed, and searched by reference number.

An example of a data template/worksheet is shown in Figure 4. In addition to a line description of each datum required, several other characteristics should be noted. The first line tells the user the

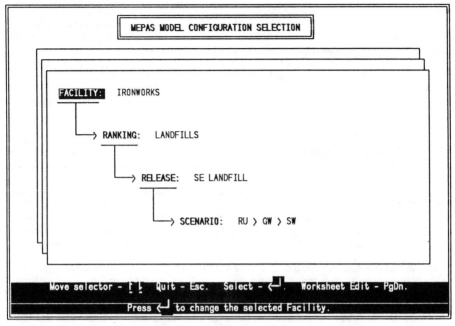

FIGURE 3. MEPAS Model Configuration Selection Menu (Create/Update Examine)

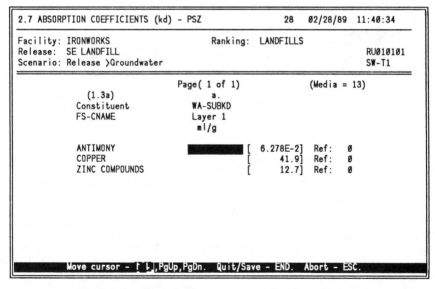

FIGURE 4. Partially Saturated Zone K_d Values

section number and name of this template, which corresponds to a section in the parameter definition guidelines. The names of the facility, ranking unit, release unit, and scenario are displayed so the user does not need to remember the current configuration.

The process of data entry to the worksheets is subject to several controls and guides designed to help eliminate data input errors.

- A range filter is used for values that must be within a specific range.

- Typical values are provided for selected input parameters to give the user an indication of values appropriate for the selected application. These typical values are normally enclosed in brackets to the left of the data entry location for the parameter as a guide for the user. For example, the shell contains algorithms for estimating typical values based on the site-specific data for the equilibrium coefficients (K_ds). Other typical values provided for the user include soils information from a soil characteristics database based on the selection of soil type for a site.

- Detailed guidance is provided for every input parameter in the supporting documentation (6). This guidance is referenced directly to the template and input parameter. The range and units required in the shell templates/worksheets are listed. Valid site-specific measured values are always the first choice for defining a parameter. If such data are not available, the user is directed to define site-specific values from other sources. The other sources either take the form of charts, graphs, or maps reproduced in the guidance, or other readily available information (e.g., regional soil maps).

- Importing climatological joint frequency wind summaries and population distribution data from standard external sources is allowed. This saves the user from having to input these large data matrices.

- When the user exits a worksheet with a quit/save, the status of that template is determined, stored, and displayed in the template selection list. A scenario cannot be included in an analysis if any of its worksheets are incomplete.

<u>Transport and Exposure Analysis</u>. When the Run/Print Reports/ Worksheets option is chosen at the main menu, a submenu is displayed with selection options that enable the user to execute the MEPAS transport and exposure codes and print results from the model run.

The first step is to create the ASCII input files required by the various MEPAS components. A user-defined name is used as a file name to identify the particular run. The configuration of a data set for a run can be changed as desired, but will include a facility, ranking unit, one or more release units, and up to 10 scenarios per release unit.

The second step is to select the Run MEPAS Model option from the Run/Print Report/Templates menu. The user selects the data sets to be

run, and the model will be invoked for each data set specified.   The
intermediate and final results are written to ASCII files.

Printing Worksheets and Analysis Results.   After the scenario configuration is complete, required templates can be printed.   These
printed templates or worksheets have several functions.   If printed
early in the cycle, they can be used as working papers to assist in
compiling the data before entering it into the shell.   They can be
printed at any time to reflect the current state of the database, and
when the analysis is complete, they can accompany the analysis documentation for quality assurance and control requirements.   The contents of the reference list can also be printed from this option.

   After the analysis is complete, any of the MEPAS model input and
output files for the analysis can be printed.   Normally, the final
summary output file will be the most useful.   Intermediate files are
provided to allow the user a means of better understanding the data
contained in the summary output file.

Uses of the MEPAS Shell

The MEPAS shell provides a framework for organizing the process of
conducting an analysis.   Before development of the shell, much effort
was expended in conducting MEPAS applications with duplication of
effort and organization, and data entry problems.   The shell helps the
user through the complex process of defining the problem, defining what
data inputs are to be required, entering these data inputs, and running
the environmental simulations.

   The MEPAS shell enables faster, more efficient, and better documented applications.   With the addition of the MEPAS shell, fewer
resources need to be directed to computer-level details of model implementation.   Because the special level of computer expertise (i.e.,
knowing how to enter data in files to match a FORTRAN format statement)
was eliminated, the time for training data entry personnel was greatly
reduced.

   The MEPAS shell has elements of a knowledge-based system.   The
source term, environmental.transport, and exposure assessment data
entries build a database of information that can be used to define
additional environmental problems.   This feature, which was added to
simplify evaluation of similar or related problems, will also be useful
in the evaluation of remedial action alternatives for site cleanup
using a baseline case.   Also the knowledge base in the constituent
database grows as MEPAS is applied to new constituents.

Summary

The development of MEPAS for application to evaluations of large numbers of environmental issues based on potential health impacts is complete.   Whether used to evaluate a single site with many environmental
impacts, a facility with multiple releases, or a collection of facilities with releases, MEPAS is an appropriate tool for screening the
relative importance of environmental issues in a scientific and objective manner.

Acknowledgments

This work was supported by the U.S. Department of Energy (DOE) under Contract DE-AC06-76RLO 1830. Pacific Northwest Laboratory is operated for DOE by Battelle Memorial Institute.

Literature Cited

1. Whelan, G.; Strenge, D. L.; Droppo, J. G. Jr.; Steelman, B. L. The Remedial Action Priority System (RAPS): Mathematical Formulations; PNL-6200, Pacific Northwest Laboratory, Richland, Washington, 1987.

2. Droppo, J. G. Jr.; Whelan, G.; Buck, J. W.; Strenge, D. L.; Hoopes, B. L.; Walter, M. B. Supplemental Mathematical Formulations: The Multimedia Environmental Pollutant Assessment System (MEPAS); PNL-7201, Pacific Northwest Laboratory, Richland, Washington, 1989.

3. U.S. Department of Energy (DOE). Environmental Survey Preliminary Summary Report of the Defense Production Facilities; DOE/EH-0072, U.S. Department of Energy, Environment, Safety, and Health, Office of Environmental Audit, Washington, D.C., 1988.

4. U.S. Environmental Protection Agency. Analysis of Alternatives to the Superfund Hazard Ranking System; Prepared by Industrial Economics, Incorporated, Cambridge, Massachusetts, 1988.

5. Droppo, J. G. Jr.; Strenge, D. L.; Buck, J. W.; Hoopes. B. L.; Brockhaus, R. D.; Walter, M. B.; Whelan, G. Multimedia Environmental Pollutant Assessment System (MEPAS) Application Guidance Volume 1 - User's Guide; PNL-7216, Pacific Northwest Laboratory, Richland, Washington, 1989.

6. Droppo, J. G. Jr.; Strenge, D. L.; Buck, J. W.; Hoopes, B. L.; Whelan, G. Multimedia Environmental Pollutant Assessment System (MEPAS) Application Guidance Volume 2 - Guidelines for Evaluating MEPAS Input Parameters; PNL-7216, Pacific Northwest Laboratory, Richland, Washington, 1989.

7. Whelan, G.; Brockhaus, R. D.; Strenge, D. L.; Droppo, J. G. Jr.; Walter, M. B.; Buck, J. W. "Application of the Remedial Action Priority System to Hazardous Waste Sites on the National Priorities List." Superfund '87, Proc. 8th Nat. Conf., 1987, pp 409-413.

8. Morris, S. C.; Meinhold, A. F. Report of Technical Support for the Hazardous Waste Remedial Action Program, on Health and Environmental Risks of Inactive Hazardous Waste Sites; BNL-42339, Brookhaven National Laboratory, Long Island, New York, 1988.

9. Whelan, G.; Strenge, D. L.; Droppo, J. G. Jr.   "The Remedial Action Priority System (RAPS): Comparison Between Simulated and Observed Environmental Contaminant Levels." Superfund '88, Proc. 9th Nat. Conf., 1988, pp 539-545.

10. Whelan, G.; Droppo, J. G. Jr.; Strenge, D. L.; Walter, M. B.; Buck, J. W. A Demonstration of the Applicability of Implementing the Enhanced Remedial Action Priority System (RAPS) at Hazardous Waste Sites; PNL-7102, Pacific Northwest Laboratory, Richland, Washington, 1989.

11. Droppo, J. G. Jr.; Buck, J. W. "Characterization of the Atmospheric Pathway at Hazardous Waste Sites." Proc. DOE Model Conf., 1988, pp 1-11.

12. Hartz, K. E.; Whelan, G. "MEPAS and RAAS Methodologies as Integrated into the RI/EA/FS Process." Superfund '88, Proc. 9th Nat. Conf., 1988, pp 295-299.

13. Poston, T. M.; Strenge, D. L. "Estimation of Sport Fish Harvest for Risk and Hazard Assessment of Environmental Contaminants." Proc. 6th Nat. Conf. Haz. Wastes and Haz. Materials, 1989, pp 118-123.

14. Buck, J. W.; Aiken, R. J. "Applications of the Multimedia Environmental Pollutant Assessment System (MEPAS)." Proc. of HAZTECH Int. Conf., 1989, pp 1-10.

15. Buck, J. W.; Hoopes, B. L.; Friedrichs, D. R. Multimedia Environmental Pollutant Assessment System (MEPAS): Getting Started with MEPAS; PNL-7126, Pacific Northwest Laboratory, Richland, Washington, 1989.

16. Strenge, D. L.; Peterson, S. R. Chemical Data Bases for the Multimedia Environmental Pollutant Assessment System (MEPAS): Version 1; PNL-7145, Pacific Northwest Laboratory, Richland, Washington, 1989.

17. Hoopes, B. L.; Buck, J. W.; Friedrichs, D. L.; Aiken, R. J. The Multimedia Environmental Pollutant Assessment System (MEPAS) User-Friendly Shell; PNL-SA-16284, Pacific Northwest Laboratory, Richland, Washington, 1988.

RECEIVED March 8, 1990

Chapter 15

The Defense Priority Model for Department of Defense Remedial Site Ranking

Judith M. Hushon

Roy F. Weston, Inc., 955 L'Enfant Plaza, S.W., Sixth Floor,
Washington, DC 20024

The Defense Priority Model (DPM) is designed to provide
an estimate of the relative potential risk to human
health and the environment from sites containing
hazardous materials. The DPM evolved from a model called
the Hazard Assessment Risk Model (HARM) developed by Oak
Ridge National Laboratory from 1984-1986 for the Air
Force. The automation of DPM was done first in KES(r)
and then in Arity Prolog(r) for use on an IBM-PC/AT class
machine. The computerized model has already become more
sophisticated than the paper model and as development
continues, it is possible to take advantage of additional
expert system features. This paper is designed as a case
study of DPM development and presents the reasons for the
choice of expert system environment and its evolution,
the current scope of the model, and planned additions
that will increase the functionality of model in the
future. The methodology used to evaluate this expert
system is also described.

Work on what is now the Defense Priority Model was initiated in 1984
when the U.S. Air Force recognized the need for a defensible
methodology for ranking for cleanup sites containing hazardous wastes.
The original work was conducted by Barnthouse and his colleagues at
Oak Ridge National Laboratory and resulted in the development of the
Hazard Assessment Risk Model (HARM).(1,2) This model was then
evaluated using comparative testing by a number of reviewers and the
results led to the incorporation of a number of changes and the
development of HARM II.(3)
    HARM II considered the toxicity and quantity of the pollutants
present, two exposure routes - surface and ground water, and human and
ecological receptors. To obtain a significant score, a source, a
pathway and a receptor all had to exist since this ensured that
exposure was a possibility.

0097–6156/90/0431–0206$06.00/0
© 1990 American Chemical Society

The Air Force determined that the model needed to be computerized to be maximally useful.   They initially considered using dBase or Lotus software for implementing the model, but then decided that expert systems technology could provide some definite advantages such as the ability to:

o   Incorporate uncertainty

o   Accommodate missing data

o   Use alternative pathways to obtain an indication of a required factor

o   Manage flow through the program so that only appropriate questions are asked of the user

o   Include expert knowledge and make this available to the user

o   Include both quantitative and qualitative data in the site scoring process.

Selection of the Programming Environment

The Air Force wanted the model to be useable by a large number of geographically diverse users so the IBM-PC/AT was selected as the logical delivery environment.   A number of expert system shells were then examined and KES(r) by Software A&E was selected.

Using this software, a prototype of the system was constructed and demonstrated.   This prototype modeled exposure from only the surface water exposure routes plus the accompanying hazard and receptor scoring.   It was possible to compile this single pathway, but when attempts were made to add a second pathway scoring to include ground water exposure, the program became too large to compile.   KES also placed severe limitations on screen management.

A decision was therefore reached that the code would be translated into prolog.   Arity Prolog(r) was selected because it is the most complete and powerful of the PC-based prolog compilers.   The effort involved in changing development environments was minimized somewhat by preparing definitions for the prolog version that recognized the KES grammar and code prepared to date.   However, using a language rather than a shell meant that a number of program segments for screen management, file access, etc. had to be written which had previously been accomplished by KES.

Development proceeded and in the Fall of 1988, a version of the code that included all of the features of HARM II was available for testing (DPM version 1.0).   At this time, the name of the model was also changed to the Defense Priority Model (DPM) and it was adopted by the Office of the Deputy Assistant Secretary of Defense (Environment) for uniform use by the armed services to rank sites for remedial action in Fiscal year-1990 (4).

Evaluation

The comparative testing of the automated DPM version 1.0 was done by a team of scientists including two each of life scientists, geologists, and environmental engineers. Five sites at each of three military installations were selected and copies of the appropriate Installation Restoration Program (IRP) Phase II reports (Remedial Investigations/Feasibility Studies) were made available to the reviewers along with a copy of the current user's manual and a disk containing the code. Each site was scored by two reviewers representing different disciplines. The evaluators were given a one-day course in DPM and hotline assistance was provided during the evaluation period. The scores were then collected and summary tables prepared.

When differences in scoring were noted between two evaluators, they were questioned as to why each had made his/her selection. Several sources of differences were identified:

o   One reviewer used his disciplinary knowledge to interpret the available data and reach a conclusion while the other did not have this special knowledge.

o   Errors were made converting units.

o   Definitions were not as clear as they might have been and different conclusions were reached by two evaluators as to how a variable should be scored.

o   Reviewers differed in their reading of map distances.

Additionally, the evaluators made suggestions on how to improve the functionality of the program to make it more user friendly.

The agreement among the initial reviewers was moderately good, with the sites ranking in the same order, but the individual scores varied by more than was felt to be desirable. This prompted a number of changes in definitions and program structure. For example, two scorers looked at a site map and one felt that the distance to a well was 2.9 miles (which gave a score of 1) and the other felt that the distance was 3 miles (which was outside the scoring boundary and was given a score of 0). As a result, the distance from a target to a receptor was modified to include a matrix of population size by distance from the site to the well.

In addition to these changes, several additional major additions were made to the model at the suggestion of outside reviewers:

o   An air/soil pathway was added

o   Some exposure distances were extended by introducing matrices for scoring so that the score was a function of the number of people as well as the distance from the site

o   Automatic look-up of meteorology data for Air Force bases was added.

The first addition was a major one which involved significant research to determine how best to deal with volatiles and air/dust emissions. These additions were fully implemented in the version of DPM that was released for evaluation in March 1989.

Additionally, a number of new features were included in the computerized version to make it more functional(5); these include the ability to:

o   Answer a question one time even if it is used in several separate pathways and calculations

o   Record certainty of input data

o   Automatically convert units

o   Use alternate data if information are missing

o   Check range of input data, and

o   Change responses and to rapidly recalculate a final score

The automated version can also generate a report that includes, in addition to the scores, full documentation of the final score through comments and the certainty indication. Additionally, the automated version controls the user's passage through the model and only presents those requests for information that are deemed necessary depending on previously supplied answers.

Overview of the Model Structure

DPM considers the hazards associated with source materials, pathways that may result in exposure, and the presence of potential receptors.(6) There are three pathways in DPM:

o   Surface water

o   Ground water

o   Air/Soil (considers vaporized compounds and dust).

DPM considers both human and environmental receptors, though the human receptors are more highly weighted. The environmental receptors include both aquatic and terrestrial populations as appropriate.
Figure 1 demonstrates how the various pathway scores are combined to yield the six pathway/receptor scores per site. These six scores are then combined using a root mean square methodology to obtain a single site score (see Figure 2). All scores are normalized so that they range from 0 to 100; this score, by itself, has no meaning and should not be compared to the Hazard Ranking System (HRS) ranking number for inclusion on the National Priority List (NPL). Most sites evaluated to date scored in the 20 to 30 range, but sites have scored as high as 89 and as low as 3 so a broad range of values can be expected.(7)

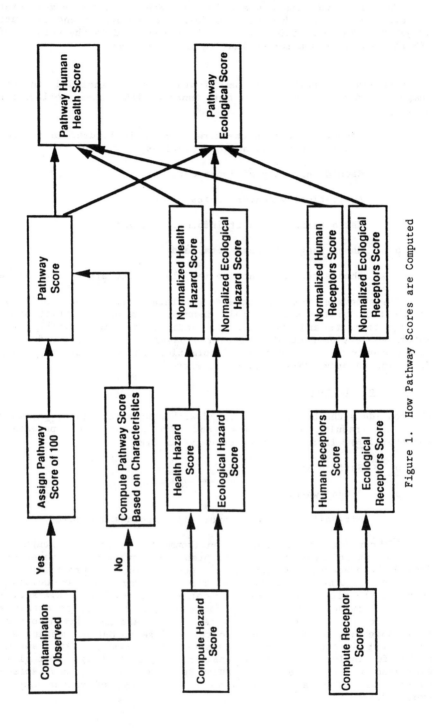

Figure 1. How Pathway Scores are Computed

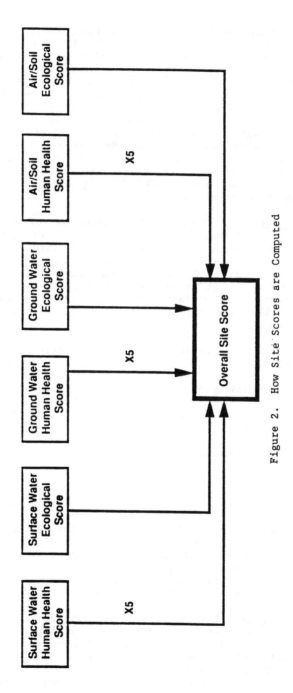

Figure 2. How Site Scores are Computed

Pathways

To better understand what is included in the pathway scores, it is necessary to examine each pathway more closely with regard to the types of data that are included. Each pathway score is computed by scoring a number of related factors; different factors have different weights. The factors correspond to individual measureable related variables such as flooding potential and net precipitation as components of the surface water pathway.. The approach used is to obtain a score for each variable and to multiply this score by a preestablished weight. The weighted scores for all factors in a pathway are then added and divided by the maximum possible score to obtain a normalized value. For each of the pathways, if a chemical release is observed in that pathway, a maximum score is assigned. However, this score can be modified by a weighting based on how well the waste/hazard is contained.

Surface Water Pathway. The surface water pathway of DPM rates the potential for contaminants from a waste site to enter surface waters via overland flow routes, or from ground water discharge to surface water. If pollutants are not directly observed in surface water, but are present in sediments or soil, there is a potential for surface water contamination so it is necessary to collect information from which their likelihood to reach a receptor can be estimated. The following variables are scored to provide an indication of this exposure potential:

o   distance to nearest surface water (scores are assigned up to a mile)

o   net precipitation

o   surface erosion potential (combination of slope and particle size)

o   rainfall intensity

o   surface permeability

o   flooding potential (location within floodplain)

The most heavily weighted factor is flooding potential with net precipitation receiving the least weight. The containment of the waste is also estimated by the scorer based on guidance and becomes an important weighting factor. Containment is a function of the type of site (e.g., old landfill, fire training area, lagoon) and the efficiency of present pollution control measures.

Ground Water Pathway. The ground water pathway ranks the potential for pollutant exposure to occur from contaminated ground water. If actual ground water contamination has not been detected, but there is contamination in soil or surface water, there is a potential for ground water contamination to occur in the future. The following factors are scored to obtain a ground water pathway score:

o   depth to seasonal high ground water

o   permeability of the unsaturated zone

o   infiltration potential (measured from net precipitation and
       the form of the waste)

o   potential for discrete features in the unsaturated zone to
       "short circuit" the pathway to the water table.

Waste containment effectiveness is also a weighting factor on the
pathway score. Of the above factors, the depth to the seasonal high
water table is the most heavily weighted factor.

Air/Soil Pathway. The original HARM model did not have an air/soil
pathway and thus could not account adequately for exposure due to
volatization of organics from soil or surface water, or for exposure
to contaminated dust. The factors that are considered in scoring this
pathway are:

o   average temperature

o   net precipitation

o   wind velocity

o   soil porosity

o   days per year with significant precipitation

o   site activity.

All of these factors are weighted evenly. A factor for waste
containment is also selected by the user and employed to modify the
final score.

Contaminant Hazards

The contaminant hazard component of DPM separately rates human health
and ecological hazards of identified or suspected contaminants in each
of the three pathways. Hazard scores are calculated differently
depending on whether environmental contamination has been detected.
For media in which contamination has been detected, health hazard
scoring is based on the concept of an acceptable daily intake (ADI).
The highest concentration regardless of the toxicity of the
contaminants observed at a site is used. The observed concentraton
is first converted to a daily intake in (ug/day) and then divided by
the appropriate benchmark concentrations (provided in the manual or
on the computer system) which are estimated ADI's. Ecological hazard
scoring for observed contaminants is similar, although an ecological
benchmark is used instead. The sum of the ecological hazard quotients
(concentration divided by the benchmark) is used for all detected
contaminants.

For media in which contamination has not been detected, health hazard scores are assigned based on the ADIs and bioaccumulation factors of contaminants known to be present at the site being rated. In this case, the score is based on the most toxic contaminant.

Scoring is similar for all pathways, though the appropriate benchmarks will vary. For example, if the pathway is surface or ground water, aquatic benchmarks will be used as well as terrestrial benchmarks. For the air/soil pathway, however, only terrestrial benchmarks are employed.

Receptors Scoring

The receptors portion of the DPM methodology rates the potential for human and ecological populations to be exposed to contaminants from a waste site. The potential receptors are considered separately for each pathway and for human and ecological targets.

Human Receptors for Surface Water. The following factors are scored to obtain a measure of human exposure to surface water pollution:

o   size of population obtaining drinking water from potentially affected downslope/downstream surface waters (up to 5 mile radius)

o   water use of the nearest surface water

o   population within 1500 feet of the site

o   distance to the installation boundary

o   land use and zoning within 2 miles of site

The first two factors listed above are weighted most heavily.

Human Receptors for Ground Water. The following factors are used as indicators of potential human receptor exposure to contaminants suspected in ground water:

o   estimated mean ground water travel time from waste location to nearest downgradient water suply well(s)

o   estimated mean ground water travel time from current waste site to any downgradient surface water body that supplies water for domestic use or for food chain agriculture

o   ground water use of the uppermost aquifer

o   size of population potentially at risk from ground water contamination

o   population within 1000 feet of the site

o   distance to the nearest installation boundary

Of these factors, the estimated ground water travel time is considered most important with the water use of the uppermost aquifer being important as well.

Human Receptors for Air/Soil. The following factors are used as measures of the potential for human exposure:

o   size of population near the site (4 mile radius)

o   land use in vicinity of the site

o   distance to nearest installation boundary

Land use has the most pronounced impact on the final score.

Ecological Receptors - All Pathways. Exposure of potential ecological receptors is determined by whether there are sensitive environments (i.e., wetlands or habitats of endangered species) within 2 miles of the site and whether there are critical environments (i.e., lands or waters specifically recognized or managed by federal, state, or local government agencies or private organizations as rare unique, unusually sensitive, or important natural resources).

Combining Pathway Scores to Obtain a Final Site Score

The scores for each pathway are obtained by combining the information on the pathway and the hazards for health and ecological receptors. The result are six subscores, one for each receptor/pathway combination. These scores are then combined using a root mean square methodology with the human health scores weighted five times heavier than the ecological scores. The final score is then normalized by dividing by the maximum possible score to obtain a site score ranging from 0 to 100.

Future Directions

In November of 1987, the DoD proposed the use of the DPM for prioritizing remedial actions and announced a public comment period in the Federal Register (4). Comments on the model were received from the Environmental Protection Agency and three states. These were considered and based on them, changes were made to the model. Fiscal Year 1990 was the first year the model was applied to DoD installations. Over 120 Army, Navy, Air Force and Defense Logistics Agency representatives were trained in operation of the automated model. Feedback is also being elicited from them as to how the model can be improved to facilitate future scoring.

Some areas that have already been identified for improvement include:

o   Logical checking of related answers.

o   Addition of materials to the chemicals files that have been identified at IRP sites but are currently missing. This is expected to include explosives and radioactives as well as individual compounds.

o   More ability to move around through the model.

o   Computation of an overall certainty score.

Work is progressing on DPM and the Fiscal Year 1990 application
will create a large body of data on actual sites. These data will be
analyzed and improvements will be made to the model as appropriate.

There is also a plan to convene a group of experts to validate the
model.

Acknowledgment

Work on DPM was conducted under contracts with the US Air Force's
Occupational and Environmental Health Laboratory located at Brooks
AFB, TX. The Technical Program Managers for this work were Capt. Art
Kaminski and Mr. Phil Hunter.

References

1.   Barnthouse, L.W., J.E. Breck, T.D. Jones,S.R. Kraemer, E.D.Smith
     and G.W. Suter II, Development and demonstration of a hazard
     assessment rating methodology for Phase II of the Installation
     Restoration Program, ORNL/TM-9857, Oak Ridge National Laboratory,
     Oak Ridge, TN, 1986.

2.   Barnthouse, L.W., J.E. Breck, G.W. Suter II, T.D. Jones, C.
     Easterly, L. Glass, B.A. Owen and A.P. Watson, Relative toxicity
     estimates and bioaccumulation factors in the Defense Priority
     Model, ORNL-6416, Oak Ridge National Laboratory, Oak Ridge, TN,
     1986.

3.   Smith, E.D. and L.W. Barnthouse, User's Manual for the Defense
     Priority Model, ORNL-6411, Oak Ridge National Laboratory, Oak
     Ridge, TN, 1986.

4.   Federal Register, Vol. 52, No. 222, p. 44304-5, November 1987.

5.   Hushon, J.M., G.M. Mikroudis, and C. Subramanian, Automated
     Defense Priority Model User's Manual, Version 2.0, Roy F. Weston,
     Inc., Washington, DC, June 1989.

6.   Hushon, J.M., G.M. Mikroudis, and C. Subramanian, Defense
     Priority Model User's Manual, Version 2.0, Roy F. Weston, Inc.,
     Washington, DC, June 1989.

7.   Hushon, J.M., G.M. Mikroudis, and N. Pandit, Final Report on
     Phase I of DPM, Roy F. Weston, Inc., Washington, DC, December
     1988.

RECEIVED April 20, 1990

Chapter 16

The Future of Expert Systems in the Environmental Protection Agency

Daniel Greathouse[1] and James Decker[2]

[1]Risk Reduction Engineering Laboratory, Environmental Protection Agency, Cincinnati, OH 45268
[2]Applied Technology Division, Computer Sciences Corporation, Center Hill Laboratory, Cincinnati, OH 45268

As in other organizations, the history of expert systems in the Environmental Protection Agency is very short. Approximately five years ago the focus of our activities was to assess the feasibility and utility of using expert systems as environmental decision aids. Last year the Agency approved a five year funding initiative to support development of a set of systems to assist in management and implementation of Superfund activities. Whereas initial systems were limited to a few engineering and technical issues, the scope of current systems includes legal, regulatory, and administrative functions. Notwithstanding this evolution in scope and funding, expert systems are not yet mainstream decision making tools in the Agency. Many decision makers are either not familiar with expert systems or are skeptical that they can provide meaningful and reliable advice. Because expert systems are a relatively new technology in practice, and inasmuch as a widely used application has not yet been developed for use in the Agency's regulatory environment, Agency use of knowledge-based systems in the future is uncertain. This paper proposes a scenario for evolution of the development environment for expert systems within the Environmental Protection Agency.

Defined generally, expert or knowledge systems are computer programs designed to provide advice concerning particular and usually specialized issues. Using rules and control strategies in conjunction with user-supplied data, they function to provide analytic assistance and consultation by logically interrelating information within a restricted domain.

As is the case in most large organizations, the history of expert systems within the Environmental Protection Agency (EPA) is brief.

0097–6156/90/0431–0217$06.00/0

During the past five years or so, substantial changes have occurred
to reduce the cost and increase the reliability and availability of
commercial knowledge system building tools. Within the Agency there
has been a corresponding increase in the level of interest in expert
system technologies. The scope of Agency decisional functions for
which such systems are either being considered or designed has also
expanded. Early EPA applications were largely confined to information and decision issues of specialized interest to their developers.
Engineering knowledge systems were developed by engineers and other
persons involved with engineering methodologies, modeling systems were
designed by model theorists and statisticians, and initiatives for
regulatory systems came from program office staff. These first
systems were conceived primarily as tools for the facilitation of
information transfer from a developer's organization to decision
makers in other parts of the Agency. As such, they were driven more
by needs as envisioned by system developers than by the expressed
needs of potential or actual end users.

   Experience has shown this approach to be undesirable for several
reasons. Development resources, which are typically substantial, are
used inefficiently inasmuch as excessive time is used for development
relative to the benefits achieved when delivery is targeted toward a
small user group. Similarly, systems can be too constrained in scope
so that the functions which they address change in nature or in
significance near the time when systems are completed. Development
costs are rarely recovered under these circumstances. These alloca-
tion inefficiencies unfortunately tend to jeopardize and impede
acceptance of the technology of expert systems. Receptiveness to the
technology itself may decline in effect, due to injudicious applications. In this environment, initial successes - systems that result
in significant decision efficiencies - are critical to realization of
the long term gains which we believe, can be achieved through knowledge systems. Largely in response to problems encountered through
these earlier development efforts, a more organized and selective
approach to assessing potential systems has been recently pursued.

   Roughly two years ago the EPA Risk Reduction Engineering
Laboratory and the Office of Solid Waste and Emergency conducted an
assessment of potential expert system applications. Approximately two
hundred persons were interviewed. The assessment group included
upper-level managers and their subordinates in EPA headquarters,
personnel from several EPA regional offices, and personnel from one
state. Criteria for selecting candidate systems included estimation
of potential resource savings weighed against estimates of development
cost, the availability of necessary expertise, long term needs (to
anticipate system service spans), and the degree to which expressions
of need were widespread. Respondents were asked to identify problem
areas most likely to impede successful completion of their decision-
making responsibilities two or three years into the future.

   Upper levels of management were contacted first at each location.
The primary intention of this approach was to solicit management
support and cooperation for future activities and to identify a
management-based perspective of future needs. Lower level management
and technical staff members were also interviewed to identify more
specific issues that could be addressed and to solicit their input
into our development efforts. In our view, involvement of all levels
of management and technical personnel was critical to the success of

any subsequent system development efforts.   The reason for inquiring about future rather than present decision support needs was to provide adequate time for the development of appropriate products and to selectively focus consideration of long term needs.   Through this process, forty decision areas were identified as potential applications for expert systems and eleven areas were selected as targets. In response to a proposal formulated by the EPA Hazardous Waste Research Committee, an initiative to support development of systems to aid in decisional processes in these eleven areas was approved. Development activities are scheduled to occur over a five year span.

The initiative has motivated a broad shift in focus for the Laboratory's system design and production work.   Whereas prior projects were undertaken to support the EPA efforts to review permit applications for hazardous waste land disposal sites; future systems will address needs in the implementation and management of clean up activities at Superfund sites.   New systems will also address a more encompassing range of issues than previous systems.   In addition to issues of parochial interest to the Laboratory advice about legal, regulatory, and administrative issues is to be provided.

Notwithstanding increases in both scope and funding for expert systems, these systems are not presently mainstream decision making tools within the Agency.   Many decision makers are unfamiliar either with expert systems as practical tools or with abstract expert systems concepts.   Likely there is significant skepticism about the capabilities of such systems to provide meaningful and reliable advice.   Moreover, the value of knowledge systems technology has yet to be conclusively demonstrated through implementation of any wide-ranging application in the regulatory realm of the Agency.

These circumstances define the current environment for our work. Presently, there is much uncertainty about the future role of knowledge engineering technologies within the Agency.   The remainder of this paper explores some of the problems associated with the introduction of expert system technologies, and offers a scenario for proliferation of the technology within the Agency.

Discussion

As with most new technologies, there are various obstacles which impede the general acceptance of expert systems.   Some managers are not familiar with the capabilities of modern microcomputers and a greater number are almost certainly unfamiliar with the essential characteristics of microcomputer-based knowledge engineering tools. Although expert system shells (commercial software tools used for building expert systems) now are largely based on a well-defined set of algorithms, to persons unfamiliar with these tools,   the concept of 'knowledge engineering' may still convey vague and even threatening connotations.   Some misunderstanding and doubt may be fostered by philosophic concerns and by particular aspects of the history of academic research in artificial intelligence.   Shortcomings in the processes of system planning and life-cycle management previously discussed can also slow acceptance of the technology.   Understanding and acceptance of expert system methods may to some extent, have even been impeded by the (perhaps too optimistic) claims of the technology's proponents and by implications from promotional literature through which some expert system tools have been marketed. Confusion about what expert systems can and cannot do naturally results.

The specialized languages (list processing languages such as LISP and languages with object-oriented data structure schemes) associated with many expert system tools and applications also tend to make the discipline of 'knowledge engineering' more obscure. Although many managers and technical professionals are now comfortable both with the procedural logic which is commonly represented by linear flowcharts and with other well-established computer-modeling practices, far fewer are comfortable or familiar with data structures concepts such as list processing procedures, frames, 'backward chaining', or truth-maintenance. The differences between these approaches to representing and manipulating information of course, produce the distinctive characteristics of expert systems. Unfortunately, such differences also tend to place expert systems technologies in a category apart from what is thought of as mainstream computing which includes data-base processing and statistical modeling, among other areas.

Lastly, as a product of artificial intelligence research within the broader fields of cognitive and computer sciences, expert system methods absorb some criticisms which have been expressed most often in philosophical terms related to these parent fields. Occasionally, such critical views have assumed that expert system inference processes cannot be trusted because the processes underlying expert human judgment cannot in principle be duplicated by a machine, but this criticism is too general to be useful. Indeed, it may well be found that some aspects of expert decision-making are not amenable to simulation or software reproduction; and inasmuch as academic attempts to code 'common sense' knowledge have largely been unsuccessful, (perhaps due to the vagueness of the concept of common sense), there is a viable basis for criticism. Yet on the other hand, the success of some expert system applications - particularly in medical and engineering disciplines - demonstrates that certain specific and essential aspects of complex judgments processes are reproducible in rule-based models.

As a practical matter it is important to consider why some knowledge systems have more utility or more expertise than others, and why some systems are more accepted than others. It appears that the most specialized areas of knowledge are precisely the areas most amenable to expert system representations. Understanding in such areas is itself more precise, and so it is possible to more precisely model such information. These considerations appropriately play a significant role in the processes of defining and selecting areas for expert system application. Aside from particular applications and philosophical concerns however, the simple fact that expert systems have already become highly effective decision aids in certain specific areas must be recognized.

For example, there are numerous military and defense-related systems, including administrative systems which assist in formulation of acquisition strategies and the development of relevant documents, battle strategy systems, and hardware maintenance advisors. Other successful administrative applications include a system used by the IRS for selection of audit candidate tax returns, and a credit evaluation system used by American Express. Successful systems in engineering areas include the XCON system which assists in configuration design of DEC (Digital Equipment Corporation) computer systems, a production scheduling system used by Westinghouse Corporation, and an engineering cost estimation system used by Navistar Corporation. Successful environmental systems include an emergency response advisor (CAMEO), and a chemical reporting aid to assist in compliance with Title III requirements (Dupont).

Future Directions

Because the technology has been successfully applied in so many fields and endeavors, we believe that continuing exploration and expansion of the scope of Agency applications is desirable. Likewise, although there are formidable special difficulties presented in the areas of design, distribution, and life-cycle management for expert systems, we believe that continuing exploration and expansion of the scope of Agency applications is imperative for efficient satisfaction of the essential missions of the Agency. Moreover, in the long term we anticipate that knowledge-based systems will become integral and widely used tools in almost all aspects of the Agency's regulatory functions. Similarly, we expect that applications will not be limited to a few high-priority areas, but that routine functions as well as highly specialized knowledge-intensive domains will be impacted by these systems.

Three major trends in the commercial world of expert systems will hasten these events. The first of these is the current growth in the availability of low-priced shell tools. Although the cost of most mainframe and workstation tools for large systems remains above $8000, the capabilities of lower price tools ($400 to $2000) are rapidly expanding. This will likely encourage competitive pricing for the larger system tools and encourage interested parties toward further experimentation. The second trend is toward the standardization of a set of core control features or capabilities among systems. Capabilities for object-oriented control, types of inferential logic or 'chaining' search processes, property inheritance schemes, and frame-based reasoning are emerging as necessary minimal features for acceptable expert system building tools. Lastly, it is quite likely that some conventional software system products will include particular methodologies from the field of artificial intelligence as enhancements to their base products. In this area, for example, the SAS Institute, a developer of statistical analysis software, is currently engaged in test projects to incorporate artificial intelligence methodologies into easier-to-use system interfaces for particular parts of its product group. In the near future, some database products may also be expected to include object-oriented access features.

Overall, the pattern of acceptance and use of expert systems will probably parallel events that occurred in more established application areas such as word processing.

Some of the basic features of word processing such as document memory and letter erasure started appearing on typewriters a few years ago. More advanced features such as font selection and text transfer were in 1984, generally not available with word processing packages. Gradually however, more of these features became available for general use. Today of course, most word-processing packages have capabilities which were formerly limited to specialized publishing systems. As a result of this downward spread of advanced capabilities from specialized software to common off-the-shelf software, sophisticated document processing techniques are now widely employed in the regular activities of most large organizations, including the Environmental Protection Agency.

Factors similar to those that shaped the change in the nature of common document production work from the functions associated with typewriters, to those associated today with desktop publishing systems, appear to be at work in the maturation of expert systems tools. Significant advances are occurring in the capabilities of commonly available tools, in the ease-of-use characteristics of these tools, and in the platforms which use these tools. The cost of both powerful processors and shells is now low enough that it is practical to implement systems with hundreds of rules and sophisticated control structures on low cost microcomputers. Graphic interfaces and more natural syntactic conventions are being implemented in available shells in order to reduce the time required for training-to-proficiency and development work.

Concurrent with these technological changes, are changes in Agency policies and staffing characteristics which argue for the effective expansion of expert systems development. Workloads are continuing to increase in all areas of the Agency without commensurate increases in personnel. In turn, the required levels of productivity in decision-making tasks are pressured. This increases the potential utility of expert systems.

Other factors also point toward the utility of expert systems for effective task management. Most of the decisions that involve implementation of Agency guidance and regulations are made by officials located within the ten regional offices and/or state offices. This distribution of decision responsibilities increases the need for methods which encourage decisional consistency. Transferring the latest information concerning new technologies, regulations, policies, and guidance to those who need it is a difficult but critical function. This too is facilitated by increases in distribution and application scope of expert systems. Lastly, in some parts of the Agency such as the Superfund Program, the personnel turnover rate is quite high. More effective and efficient methods are therefore required to facilitate new employee training. Inasmuch as many expert systems serve as training tools as well as expert decision aids, this factor also should increase the demand for widespread use of expert systems in the Agency.

Assuming that expert system use will increase within the Agency, issues related to the control and management of these systems assume greater importance. It will become necessary to determine who will develop application systems and who will support system maintenance. Delivery platforms will also need further evaluation. Effective management policies will undoubtedly evolve over time with the spread of experience with expert applications.

Through this process it is to be expected that initially, most application production work will be controlled by specialist groups dedicated to expert system development. Inasmuch as most expert knowledge sources are not currently familiar with the structural characteristics of expert systems, such persons at first will probably only be involved in the role of providing knowledge content. But once use and exposure is sufficiently increased, the potential of expert systems becomes more fully appreciated, and the tasks considered for knowledge system representation begins to include a few routine activities, then people with other skills will become involved in detailed and particular aspects of system development. Such an enhanced role will also be motivated by maintenance requirements for existing systems. Such increased involvement by persons from fields other than knowledge engineering should be encouraged.

We expect that within the next ten years expert system software and applications will become (at least nearly) as prevalent as database, spreadsheet, and word processing applications are now. Most expert-style applications will eventually be developed and delivered on microcomputer platforms. As the capabilities of microcomputer approach the capabilities of workstations, the need for more specialized machines will be reduced and this also will increase the involvement of persons with varied skills to contribute.

RECEIVED January 9, 1990

Author Index

Affiliation Index

Subject Index